Building Materials

Building Materials
Material Theory and the Architectural Specification

KATIE LLOYD THOMAS

BLOOMSBURY VISUAL ARTS
LONDON • NEW YORK • OXFORD • NEW DELHI • SYDNEY

BLOOMSBURY VISUAL ARTS
Bloomsbury Publishing Plc
50 Bedford Square, London, WC1B 3DP, UK
1385 Broadway, New York, NY 10018, USA
29 Earlsfort Terrace, Dublin 2, Ireland

BLOOMSBURY, BLOOMSBURY VISUAL ARTS and the Diana logo are trademarks of
Bloomsbury Publishing Plc

First published in Great Britain 2022
Paperback edition first published in 2023

Copyright © Katie Lloyd Thomas, 2022

Katie Lloyd Thomas has asserted her right under the Copyright, Designs and Patents Act, 1988, to be identified as Author of this work.

For legal purposes the Acknowledgements on pp. xi–xii constitute an extension of this copyright page.

Cover design by Namkwan Cho
Cover image © Getty Images

All rights reserved. No part of this publication may be reproduced or transmitted in any form or by any means, electronic or mechanical, including photocopying, recording, or any information storage or retrieval system, without prior permission in writing from the publishers.

Bloomsbury Publishing Plc does not have any control over, or responsibility for, any third-party websites referred to or in this book. All internet addresses given in this book were correct at the time of going to press. The author and publisher regret any inconvenience caused if addresses have changed or sites have ceased to exist, but can accept no responsibility for any such changes.

A catalogue record for this book is available from the British Library.

Library of Congress Cataloging-in-Publication Data

Names: Lloyd Thomas, Katie, author.
Title: Building materials : material theory and the architectural specification / Katie Lloyd Thomas.
Description: London ; New York : Bloomsbury Visual Arts, 2022. | Includes bibliographical references and index.
Identifiers: LCCN 2021023328 (print) | LCCN 2021023329 (ebook) | ISBN 9781350176249 (epub) | ISBN 9781350176225 (hardback) | ISBN 9781350176232 (pdf) | ISBN 9781350176256 | ISBN 9781350277830 (paperback)
Subjects: LCSH: Building materials. | Architecture–Philosophy.
Classification: LCC TA403.6 (ebook) | LCC TA403.6 .L59 2022 (print) | DDC 620.1/1—dc23
LC record available at https://lccn.loc.gov/2021023328

ISBN: HB: 978-1-3501-7622-5
PB: 978-1-3502-7783-0
ePDF: 978-1-3501-7623-2
eBook: 978-1-3501-7624-9

Typeset by RefineCatch Limited, Bungay, Suffolk

To find out more about our authors and books visit www.bloomsbury.com and sign up for our newsletters.

To my parents
Anne and David Lloyd Thomas
and to Skye and Christian.

Contents

List of Illustrations ix
Acknowledgements xi

1 Introduction 1

Simondon and the Specification 5
Building Materials: An ontogenetic approach 12
'Veritable Relations': Building materials as systems 15
On the Transductive Method 20

2 Specifying Building Materials 27

From Object to Process: 18th and 19th century specifications and the shift to division by trade 32
From 'Means' to 'Ends': 20th and 21st century specifications and the variety of forms of clause 44

3 Naming Materials 51

From Species to Brand Names: Changing practices of naming timber 53
Effects of Changes in Naming: The emergence of proprietary specification 59
Naming and Table 2/3: Materials as varieties of matter 66

4 Process 71

The Process-based Clause 75
'Nothing but a Transit': Hylomorphism and the forgetting of process 78
Dynamic Operations in Process-based Description 86
'Rendered Plastic by Preparation': Preliminary operations 103

5 Performance 113

Performance Specification 120

'Grounded in Such Usefulness': Material as equipment 125
'For a Given Service': 'New glass performances' 134

6 Systems of Material 153

Informed Materials 156
Simondon's 'Complete System' 160
'That Constitutive Seam' 166

7 Going into the Mould 183

Preliminary Operations 184
The Technical Object 189
Inventive Relations 197

Notes 205
Bibliography 233
Index 245

Illustrations

2.1 Articles of Agreement for a town house for Sir William Heathcote at St James Square (1734–6), RIBA Archives, HeW/1/1/2. Source: Courtesy of the Royal Institute of British Architects Library Drawings and Archives Collection.   33
2.2 Plumber's Pamphlet, Specification of Works for Newgate Gaol and the Sessions House at the Old Bailey, London (1770). Source: Courtesy of the Royal Institute of British Architects Library Drawings and Archives Collection.   36
2.3 Contents page, *Specification*, 1898. Source: Courtesy of EMAP.   41
2.4 Joiner title block, *Specification*, 1898. Source: Courtesy of EMAP.   42
2.5 Front Cover, National Building Specification, 1973. Source: Courtesy of NBS.   46
3.1 Table 2/3, *Construction Indexing Manual*, RIBA Publications (reprinted 1969). Source: Courtesy of NBS.   68
4.1 Preparatory sketch for Wall One showing jig construction, Alan Chandler (2004). Source: Courtesy of Alan Chandler.   88
4.2 Preparatory sketch for Wall One showing concrete pour and tensioning process, Alan Chandler (2004). Source: Courtesy of Alan Chandler.   89
4.3 Specification for the Works, Elfrida Rathbone School for the Educationally Subnormal, (1961). Architect – John Bancroft with London County Council. RIBA Archives, LCC/AD/1. Source: Courtesy of the Royal Institute of British Architects Library Drawings and Archives Collection and London Metropolitan Archives, City of London.   94
4.4 Specification of Works for a House at Farnham Common, Bucks. (1934). Architect – Val Harding with Tecton. RIBA Archives, SaG/17/3. Source: Courtesy of the Royal Institute of British Architects Library Drawings and Archives Collection.   97
5.1 Pendulum test being undertaken at CERAM Research Ltd. Photographs Kevan Brassington. Source: Courtesy of NBS.   123
5.2 DIN Ramp text being undertaken at CERAM Research Ltd. from *NBS Journal* 14, May 2009. Photographs Kevan Brassington. Source: Courtesy of NBS.   124

5.3	Illustration of Pilkington glass coatings described 'for' their performances – image taken from David Button and Brian Pye (eds) *Glass in Building* (Oxford: Butterworth Architecture, 1993). Source: Courtesy of Nippon Sheet Glass Co., Ltd.	142
5.4	Test rig. Figure 16.3, Test rig, Pilkington Glass Ltd., St Helens, UK, in Button and Pye (eds) *Glass in Building* (Oxford: Butterworth Architecture, 1993), 258. Source: Courtesy of Nippon Sheet Glass Co., Ltd.	146
6.1	DSDHA Potter's Fields Park Pavillion, Image of high performance burner appended to specification. Source: Courtesy of DSDHA.	176
6.2	Sarah Wigglesworth Architects, Stock Orchard Street house. Schedule attached to letter from London Borough of Islington. Source: ©Sarah Wigglesworth and Jeremy Till.	177
6.3	Sarah Wigglesworth Architects, Stock Orchard Street office. Construction drawing of sandbag-filling rig. Source: ©Sarah Wigglesworth and Jeremy Till.	178
6.4	Stock Orchard Street office. Detailed section through sandbag wall showing inner performing wall. Source: ©Sarah Wigglesworth and Jeremy Till.	179
6.5	Stock Orchard Street house. Axonometric section through straw bale wall showing structural 'ladders'. Source: ©Sarah Wigglesworth and Jeremy Till.	180

Acknowledgements

This book has been a long time coming, which means there have been too many people along the way who have contributed to its genesis than can be acknowledged here. Most of all I have to thank a lovely science postdoc. Eileen was studying the disappearance of honeybees, and I met her at an away day for women at Newcastle University, entitled 'Managing Your Career', as if prospects for women would improve if only we were a bit better organised. I told her I was struggling to make any progress turning my PhD into a monograph. As a commuter I had to be away from my young child 3–4 days a week, I explained, and when the summer break came, I wanted to spend the time with them. 'I'm sorry' she told me, 'but this summer you need to get childcare.' My heart breaking, I did, and finally made a start. This manuscript has been finished in very different circumstances, under lockdown, when I have the pleasure of being with my family every day. My gratitude and love to my partner Christian and to Skye, now a teenager, for always making it possible for me to write, teach and research, and do the work I enjoy, and for bringing so much happiness to the rest of life. Without this kind of support 'managing your career' simply wouldn't ever have been possible. For 'creating the conditions' throughout this process, special thanks also to friends Donna Fitzgerald, Sonia Hibbs, Rachel Thomas, David Weston Thomas, Rona Lee, Brigid McLeer and Jean Copping, to my sister Sally and her lovely family, to my parents Anne and David, and to Christian's parents Gisela and Helmut, to Karen Watzke, Judy Lloyd Thomas, Victoria McCaffrey, Barbara Falk, Adam Falk, Jim Falk, Sue Rowley and to Michele Foux.

In this book's original life as a PhD thesis I had the great good fortune to have two exceptional supervisors, Adrian Rifkin and Peter Osborne, and the ever-stimulating and challenging environment of the Centre for Research in Modern European Philosophy, then at Middlesex University. I was very lucky there to get to know another Simondon admirer, and want to thank especially Cécile Malaspina for her patient discussions with me and for her help with aspects of translation. Specifiers who've given their time and expertise include my dear friend Finbarr Finn and his colleagues at Allies and Morrison, Astrid Lund at the NBS, Nick Schumann, Tony Brett, Deborah Saunt of DSDHA, Simon Tucker of Cotterell Verneulen, Helen Stratford of Mole Architects and

Rene Tobe. I am also grateful to Fiona Orsini and Charles Hind at the RIBA Drawings and Archives collection for all their help with resources.

My thesis examiners Andrew Barry and Mark Dorrian were both incisive and generous, and encouraged me to turn the PhD into a book. 'Do it straight away,' they said. If only I had listened! I was lucky to later become a colleague of Mark at Newcastle University, and he and my head of school Adam Sharr have supported this project and many others throughout. Injections of enthusiasm came along the way from Cynthia Davidson, and now from James Thompson and Alexander Highfield at Bloomsbury. Other mentors and interlocutors much appreciated in the preparation of this book, have included John Gelder, Peg Rawes, Julieanna Preston, Hélène Frichot, Tijana Stevanović, Catalina Mejía Moreno, Liam Ross, Brady Burroughs, Helena Mattsson, Peggy Deamer, Adrian Forty, Stephen Walker, Sarah Wigglesworth, Jane Rendell, Megha Chand Inglis, Alan Chandler, Robert Carvais, Aggregate Architectural History Collective, Silke Kapp, João Marcos de Almeida Lopes, Will Thomson, the *Capital* Reading Group and the late, wonderful Julia Dwyer, whose friendship was my great good fortune. I'm grateful also to all those who have made invitations and given platforms to present, discuss and publish aspects of this research.

Who would have guessed though, that an interest in specifications could also have cemented some long-lasting collaborations and friendships! Recently I've had the pleasure of reading and working together with Heidi Svenningsen Kajita on her project (Im) possible Instructions. But this project only came out of the library and started circulating and building a 'head of steam' when I met two superb researchers, Tilo Amhoff and Nick Beech, who were as interested in documents and the production of architecture as I was. Our work together developing two symposia, first 'Further Reading Required' (2011) and then Industries of Architecture (2014) and related publications has shaped my ideas throughout the process of reworking this book into its present form, and has been one of its best rewards.

In the prolonged period since the major part of this research was undertaken there has, happily, been exponential growth in interest in the work of Gilbert Simondon, not to mention the long-awaited translation into English of his two major works, and also in the theorising and study of materials. I have tried to include pointers where most relevant, but have not been able to do justice here to the scope of the rich and important work in these fields and all the insight it lends to my arguments. This will come in future publications. This manuscript is being completed in the quiet of my childhood home, where my parents encouraged questions and nurtured curiosity. This book is dedicated to them, sadly no longer alive to see it published. And to beloved Skye and Christian. Because based on current performance it could be quite a wait until the next book is completed!

1

Introduction

'Through most of history', claimed the historian of metallurgy Cyril Stanley Smith in a lecture given over 50 years ago, 'matter has been a concern of metaphysics more than physics, and materials of neither'.[1] Despite the 'material turn' now in full swing across so many disciplines, from political science to gender studies and also in my own discipline of architecture, Smith's claim that materials have been overlooked still deserves our attention. His complaint is with philosophy and science; disciplines he shows that have, since Aristotle, made use of experience and knowledge gained in the workshop to develop their concepts of matter, deriving their ideas from those who worked directly with 'the wonderful diversity of real materials', but left them behind in their theories. Either, as in physics or most philosophy, accounts have ignored the rich variety of materials and eliminated what Smith calls the 'funeous aspects of the world'. Named after Jorge Luis Borges's character 'Funes the Memorious' who remembered everything, 'funicity' is Smith's own term for the tendency of material structures to lock into patterns that reflect the local events through which they came into being, and thus for diverse histories to exist through these material records into the present.[2] Or, as in chemistry, efforts to study and explain matter via the atom abandon the human scale, which Smith points out is also the realm in which material structures 'have achieved something like the maximum significant degree of funicity'.[3] In other words, we find the richest variety and funicity of entities, not at micro or macro scales, but at the scale of our everyday experience.

Architecture as a discipline must necessarily engage with materials at this intermediate scale and make use of their 'wonderful diversity' in the construction and inhabitation of buildings. So it is paradoxical that it nevertheless resorts to concepts of matter, drawn from science, philosophy and fine art, as a means to conceptualize building materials. Nevertheless, the strange abstracted status of building materials as matter in architectural practice is quite easily explained. The architect emerged as a protagonist only when techniques of drawing enabled the practice of design to be separated from the building site, and from building materials. The orthographic drawing maps out geometric form and dimension, but leaves materials blank. It is left to marginal notes or other documents appended to the drawing and usually considered supplementary to it, such as specifications or bills of quantities, to provide information about materials. Such practices reproduce and lend credibility to an idea of materials as substrate for form, and as such architectural thinking has tended to align materials with an Aristotelian notion of matter. For Aristotle, all entities are always a composite of both matter and form, where matter is that which is correlative with form in any substance but cannot bring about any change. The term now used for this schema – 'hylomorphism' – reflects this necessary pairing of *morphē* or form, and *hyle* or timber, and indeed Aristotle was the first to use this everyday term for a building material to stand in for the concept of 'material in general'. Like other early philosophers of matter, as Smith shows, Aristotle derived his formulation from the ordinary practices of the workshop. The philosopher Gilbert Simondon, whose work is central to this book, also notes the dependence of this abstract schema on technical activity. The form/matter schema makes use of a process of technical production (and just one type of production; casting) for its paradigm. But, as Simondon shows, the universal application of its explanatory power to all things, including to living beings – and we could add, to architectural production – proves problematic.

One aim of this book is to show the limitations of considering materials in architecture in terms of the form/matter schema and to set out an alternative theory of building materials. I take a very particular route to do this, turning to found descriptions of timber, glass and concrete as they are to be used in construction, and to the anti-hylomorphic, process-oriented philosophy of Gilbert Simondon. But I am by no means alone in rejecting hylomorphism as an adequate means to understand matter in architecture. Both technological developments such as 3D printing, smart materials and composites, and innovative practices exploring emergent form, craft and material invention have revealed the explanatory limitations of the form/matter schema. A number of enquiries are explicit about the need for revised notions of matter. For practitioners and theorists such as Lars Spuybroek and Neri Oxman matter should be understood as active – with its own potential to participate in form-taking.[4] Following Jane Bennett's *Vibrant Matter*, Rachel Armstrong goes further in her book *Vibrant Architecture*, arguing that all matter has the capacity for liveliness, even intelligence, and that divisions between organic and inorganic matter are artificial.[5] Starting from eco-feminist positions Peg Rawes and Hélène Frichot argue that our present times demand new ecological paradigms in which distinctions between human/non-human; living/non-living and the environment are replaced with a relational approach to our shared materiality.[6] Outside architecture, the work of Maria Puig de la Bellacasa on relations of care through permaculture practices with soil, or Astrida Neimanis' work on bodies of water, are some of the most compelling studies of materials that at the same time challenge hylomorphic concept of matter.[7]

These writers do engage with concrete examples; experimenting with emergent form or working in the lab with living matter, or studying artists' works, resource extraction or stem cell research. But the commonplace world of mainstream building materials is rarely touched upon in this body of work. The New Zealand based architect/artist Julieanna Preston is a rare exception. Her performative and textual works engage with a range of materials – from the incised stone of an ancient city wall, to an even older lump of coal, cherished in the hand so we pay attention to it before letting it go up in smoke, from a submerged timber post of a pier to bales of untreated wool.[8] Preston's explorations of more prosaic building products – such as MDF and gypsum – operate through her efforts to interact with them by cutting through their bland, blank surfaces to find their rough, expressive interiors. Preston reveals the degree to which the specificities of these ubiquitous materials are normally withheld from us. Through her actions, the featurelessness of these materials, their capacity to disappear into a background invisibility, is shown to have been carefully contrived and crafted.[9]

This book takes on Smith's call to attend to materials at the scale they are used, encountered and (with engineered materials) designed, instead of moving past them to generic matter, or matter as it appears at other scales. But why ask this question in architecture? Why try to establish a theory of materials as they are used in this practice and what is at stake in seeking to challenge the form/matter schema in the case of building materials? First, one of the problems with the dominance of the hylomorphic schema is that it privileges active form over passive matter, and gives agency to form. This structure supports the prevalent idea that the architect shapes form, leaving 'matter' secondary and interchangeable, degraded with respect to the more valued form, as expressed for example in Walter Gropius's statement that, 'Matter in and of its own is dead and without character. It draws life only from the form that the creative will of the artist breathes into it.'[10] Moreover, in recent decades, the reality of practice has come to resemble the schema yet more closely, because design and build contracts can literally give over material and construction choices to the contractor, taking away the architect's responsibility for material design, and leaving them as designer of form only, or to use a term from industry, of 'visual intent' (as we will see in Chapter 6 'Systems of Material'). Over and above any principles that architects should be involved in material selection because it makes for better looking design, there are important arguments that architects need to assert their responsibility for material choices at all levels of detail, to ensure sustainable sourcing and futures for materials, and that materials used are safe, at a time when cutting corners for profit and 'efficiency' is rife (as happened in the early

19th century), but governmental regulation and monitoring of compliance has weakened.

Second, there is a problematic tendency in architectural discourse and practice to denigrate the 'technical' and pragmatic aspects of building, even to consider them as outside the discipline proper. Discursive attention to materials is acceptable if it concerns aesthetic preferences; how should the building look, how should it feel? But debates are considered merely technical, if they concern the production of the building; the manner or economics of construction, issues of labour, legality, regulation and so on. According to the Marxist historian and theorist of labour and architecture, Sérgio Ferro, the relegation of technics and labour to architecture's underside, or the privileging of design *over* the building site is no accident (in the original Portuguese, '*canteiro e o desenho*' or '*dessin/chantier*' in his concise formulation[11]). Architecture's techniques of design and project organization have been developed precisely so that the maximum surplus value can be extracted from construction labour. According to Ferro, the separation of architectural drawing from material labour on site, enables the control and exploitation of building workers. Architectural critics and historians, who reproduce the wilful neglect of production and technics, affirming that architecture is something other than, or 'over' building, are simply the ideologues of the status quo.[12] Even putting this strong claim aside, we can see there is a problem with refusing to engage with the technics of building materials. The last few decades have seen unprecedented levels of change in the production of building materials and in how they are deployed in building, in particular the widespread use of composite, smart and performance-engineered materials and new forms of contract. By simply accepting these new materials as no more than additions to the available palette of options, we fail to interrogate their effects and the extent to which they, and how they are deployed, are altering the built environment and even the concepts we use in our discourses about it.

Lastly, and by no means least, I turn to building materials because of the sheer wonder and pleasure I take in their diversity – from the many species of timber and stone with their various natural properties and beauty, that have traversed the world for centuries, to the mixing of mortars, paints and cements and the complex processes of preparing metals and glass, to the many tiny components; nails, screws, hooks, locks, hinges, levers and handles for every situation, and the proliferation of products; wall ties, drips, insulation panels, tiles, floor coverings, waterproof sheets, felts, asphalts that characterized the building industry from the late 19th century to the present day – and all this without even taking into account the diversity of materials of local building cultures around the world. It seems a travesty to reduce this display of difference and variety, this 'funicity' of past material invention and ingenuity, spread out before us (though often hidden from the eye) in the environments we inhabit, to a concept as dumb and homogenous as matter! What if architecture allowed this variety of materials to inform its material concepts, at their own scale? What would a theory of building materials look like that drew less on notions of matter from philosophy and science, and more on how materials are defined and mobilized within architecture's own practices?

Simondon and the Specification

To pursue this exploration I make two key distinctive moves in terms of theoretical framework and primary resources. The first is to develop a theory of building materials with particular reference to the philosophy of Gilbert Simondon. At the start of the research, it was only his critique of the hylomorphic schema that seemed relevant, but as the research progressed his 'process-oriented ontology'[13] and discussion of technical objects and systems became central to my theoretical approach to building materials. I will return to Simondon's work in the second part of this section and it is introduced in detail from Chapter 4 'Process' and throughout the subsequent chapters. To study building materials I look not to the substances themselves, but to a rich but neglected historical and contemporary resource of writing on and about building materials – their descriptions in architectural specifications. The history of these documents is a vast and underexplored topic, and I limit my study to examples of specifications from London and its environs, setting out a brief survey of its development since the 18th century in Chapter 2, before organizing the main body of the book around three types of material description found in the specification (Chapter 3 'Naming Materials'; Chapter 4 'Process'; Chapter 5 'Performance').

It is not immediately obvious that Simondon's philosophy and the architectural specification could be meaningfully read together. Turning to each and then to both together, as we will see in the final section of this introduction, were originally intuitive moves. Following Simondon's critique of hylomorphism I foreground the *genesis* of building materials – the processes through which they come into being – and borrow Simondon's term 'ontogenetic' to describe this approach. And in line with Simondon's pursuit of 'transductive' thought, in which problematics emerge from a domain rather than by the application of a principle from elsewhere, I refrain from justifying the focus on Simondon and the specification through recourse to an external rationale. Instead the relationship between the two emerges through the process of thinking them through together, which starts in Chapter 4 and is also sketched out in this introduction.

Although much theoretical, historical and speculative attention has been paid to the architectural drawing with its prescription of *form* in advance of building, almost nothing has been written about the specification, which makes use of written language to prescribe various aspects of building, including (but not always, and in a variety of ways) the specification of materials. No doubt this is due to the disregard architectural historians and critics have tended to have for the technical, legal and contractual aspects of building. Indeed, over the years of this research, I have sometimes met with hostility for including these documents in architectural discussions, though admittedly interest has grown during this period in the area of technical literatures and architectural 'bureaucracy' (albeit, a term that already through its negative connotations, predetermines a mode of understanding the objects it studies) and more broadly in paperwork in the field of media studies.[14] My own research, and collaborations with Tilo Amhoff and Nick Beech such as *Further Reading Required* (2011) and *Industries of Architecture* (2014, 2015) might also have contributed to this shift.

There is no singular definition of the specification. I use the term 'specification' here to refer to more than its tightly defined (although plural) manifestations in contemporary architectural practice. In the sense of the text-based 'supplement' that accompanies a set of drawings as part of a building contract, the specification first emerged in Britain in the late 18th century. But more broadly defined as a list of materials to be used in building drawn up in advance, such written descriptions greatly predate the architectural drawing, and have been much more prevalent, because of course the specification of materials is so closely related to accounting. John Gelder, whose brief but excellent survey in his book *Specifying Architecture: A Guide to Professional Practice* remains the most comprehensive history of the document, includes a wide range of examples from accounts inscribed in stone on the base of Greek temples to medieval contracts on parchment known as 'indentures' that were effectively drawings in words and paid more attention to dimensions of spaces and structure than to the materials themselves.[15] Specifications can be as simple and informal as a single sheet of foolscap with a typed list of materials and components to be erected as a sausage factory in a residential corner of London in the 1930s[16] or as detailed and formal as the bound books more than 160 pages long and written in intricately detailed prose that the London County Council's Architect's Department produced in the 1960s.[17]

Studies that engage with the documents themselves – as opposed to reading 'through' them for the information they contain – are rare.[18] Some histories of building include detailed descriptions of these documents in their accounts of particular periods. Howard Davis and John Summerson, for example look carefully at some of the specifications produced in Georgian London, in order to better understand the shift from the medieval organization of building through guilds and master craftsmen, to capitalist building production.[19] Merlijn Hurx reads specifications to show how the emergence of the modern architect influenced 15th century developments in the organization of building production in the Low Countries, as much as in Renaissance Florence.[20] Examples of specifications studied in the context of architectural design include Michael Osman's study of specifications for balloon frame construction as a means to establish hierarchies between the architect and the craftsman.[21] The specification is a primary resource in Mhairi McVicar's study of the design of the detail in 20th century architecture, *Precision in Architecture: Certainty, Ambiguity and Deviation*.[22] Read together, these studies draw our attention to the intensely local nature of the building cultures that shape the nature of the specification in any one period, which prohibits any attempt to track the broad sweep of the changing formats of specifications through history. As Michael Ball has put it:

> In order to explain the development of the construction industry, it is necessary to examine the social relations that have evolved in individual countries rather than attempt to plot a path towards the emergence of a large-scale market. This is one of the peculiarities of the construction industry. Even between advanced capitalist countries there are wide variations in the social forces influencing the industry's development.[23]

For example, while northern European countries shared the guild system whose glaziers, masons and carpenters had relative power and autonomy, the shift described by Summerson and Davis and others away from this organization of building happens almost a century earlier in the Low Countries than in Britain. In the US the specification emerges only with the proliferation of small-scale building systems across the country in the mid-19th century and is associated much earlier with standardization and building products than in Britain. Some countries, such as Australia, followed the specification practices of the UK although their building industry developed in different ways. Patterns of specification practice vary widely from region to region. Changes over time in format and content in different locations follow no singular line of development, even if there may be features that documents share in common with those from very different contexts. Therefore it is only by focusing on a localized building culture that patterns and trends in specification practice can be gleaned.

The majority of historical documents referred to in this book are held in the archives of the Royal Institute of British Architects (RIBA) and to keep the selection local it has been limited to those that specify buildings in London and the counties around the city. The contemporary specifications I look at are also for buildings within the London area. Containing their geographical context reduces the number of variables and draws into sharper focus the divergence between the organization, content and language of the earliest document in the archives for a house in St James Square (1734) and of the most recent document I include (2006) for an office building in the City of London not more than a couple of miles away. In contrast to the format of the specification, the format of the architectural drawing and its orthographic techniques has changed little across this same period, although the number of drawings produced for a building project has multiplied exponentially. The extent to which specification practices vary across the world today, and the histories of their localized developments, are important areas for future research but the breadth and complexity of such a study is well beyond the scope of this book. However, the tendencies and shifts in specification practice identified by studying the London building context, and their significance beyond the documents themselves and the pragmatics of construction and contractual administration, provides a number of questions and frameworks that are applicable to other contexts.

So a central premise here is that the form and content of the specification, and the ways that materials are described within it are subject to radical change, and emerge out of very specific cultures of building at any given time. How the documents are structured, what they describe or omit, the language in which they are written and indeed the material technologies that produce them as artefacts, cannot be separated from, to use Howard Davis's term, the specific 'building culture' that requires them. As such, specifications, in all their diverse manifestations, can be read for much more than the information they give about specific buildings or construction techniques. For the historian they track (and more visibly than in drawings), the ever-changing ways in which buildings and indeed their materials and components are made, and the ways building is organized, financed and formalized. This book includes only a very few pictures of the documents themselves. Extracts from specifications are differentiated from other quotations in the book by the use of text boxes to mark them out. Documents from particular historical periods (18th, 19th or early 20th century) or formats (London County Council, National Building Specification, Schumann Smith) are indicated by the use of a related font, and where possible, I copy closely the formatting of the original texts.

Over and above, the dazzling array of materials and building products named in these documents, and the variations in terminologies that can be traced over the centuries, what is central to the analysis and argumentation in this book is the identification of some very distinctive differences in *how* materials are described in the specification. This might mean that the same material is described in a radically different way between documents written in different historical periods, but equally, a different form of specification could be used for the same material in documents for two contemporaneous building projects. Glass for example might be described with reference to a set of material standards – 'Best seconds' or 'good thirds'– or by a name such as 'Ratcliff Crown Glass' that specifies the material by a product's manufacturer. Conversely, in the contemporary 'open' type of specification, where the selection of the material is left to the contractor to determine, the aim is to *avoid* naming the material. Instead the material is defined by the thermal, structural or acoustic behaviours it is to achieve – how it will 'perform' in the finished building. Apart from some relatively minor forms of description (such as reference and relational clauses) the three main types I identified in my own survey of documents were: those that named the material (admittedly in a variety of ways, as we will see in Chapter 3); the recipe or process-based clause that gives details of how a material is to be made up on site or worked (see Chapter 4); and the performance clause that, as we have just seen with glass, describes a material through its performance or behaviours in the finished building (Chapter 5). By studying how building materials are represented through the language of the specification rather than looking directly at the materials 'themselves', it becomes apparent that each form of clause amounts to a distinct conceptualization of a material; simply a substance in the case of naming; something that is made, or even a process, with no view to what it is for (the recipe); and as that which acts in the case of the performance clause. The intention is that by working outwards from these conceptualizations of materials that emerge from within architecture's own practices and by being sensitive to the differences between them, a more nuanced theory of materials as they are conceived and used in architecture can be developed than if one overriding concept of matter from another discipline (such as science, philosophy and art) was simply translated across and applied to building materials (Chapters 6 and 7).

My curiosity about these 'forms of clause' was initially little more than a hunch, half-remembered from my days in practice writing specifications for small projects more than 20 years earlier. I had initially been looking at artworks and reading Jacques Derrida, Julia Kristeva, Roland Barthes, Jennifer Bloomer and the poetry of Francis Ponge, Ron Siliman, Charles Bernstein and others as means to study the language of materials, and was not expecting to engage with histories of building, architectural practice and contractual organization.[24] But once I embarked on the archival work necessary to track the instances and histories of these distinct forms of clause (and indeed through that survey, confirmed my hunch), it became evident that the logics of their appearance can only be understood with reference to the highly localized building practices that produce and make use of them. For each form of clause different manufacturing, regulatory or contractual regimes determine the mode of description. How the material is described in the document is evidence of the building culture that produces the language and terms of its mode of specification. The form of clause used also varies between types of project, depending for example, whether a private individual a government body is paying for the building, whether the material techniques are new and innovative, or familiar, even irrelevant, or to what extent the architects need to defend themselves against litigation. Thus, the decision to look at building materials via the architectural specification, and moreover through the forms of clause, necessarily grounds the theory of materials developed here in the pragmatics and technics of building production, that for some critics and historians would render it not an *architectural* theory of materials at all. Contra that position, and in fact a discovery made only through the process of this research, I argue against the notion of an architectural discourse that could be hived off and separated out from its 'constitutive seam' as so Frederic Jameson has so aptly phrased it (Chapter 6). By bringing this document used and written by architects but also written off as 'merely technical', 'merely legal' into a theoretical study, I want to show that without fully determining them, the economics and pragmatics of building condition our aesthetic discourses, our designs and the concepts we use. This is a suggestion Zeynep Çelik Alexander also makes with respect to the influence of mechanized scanning techniques on concepts of form.[25] In the case of building materials we are better equipped by knowing these processes than by ignoring them.

The second move is more counterintuitive and also started with a hunch. To develop this theory of building materials I turn to the process-oriented philosophy of Gilbert Simondon whose relevant arguments are introduced in detail and step by step, in Chapters 4, 5, 6 and 7, in tandem with discussions of the specification. Interest in Simondon's work outside France has been growing rapidly since its influence on Gilles Deleuze's ideas of individuation became well known, but he is better known in the English-speaking world as a philosopher of technology. His major work on technology (and his minor thesis) *Du mode d'existence des object techniques* was originally published in French in 1958 and has recently been translated into English as *On the Mode of Existence of Technical Objects* by Cécile Malaspina and Jon Rogove.[26] I refer to this book throughout as *Du Mode*.

Simondon's major work on individuation *L'individuation à la lumière des notions de forme et d'information* was submitted as his doctoral thesis in 1958, but was only published in full in 2005.[27] An English translation by Taylor Adkins of the full work, *Individuation in Light of Notions of Form and Information*, and based on the 2005 edition, with a second volume of supplementary texts, finally became available in 2020, not quite in time for the preparation of this book.[28] In fact, my primary reference for Simondon's philosophy of individuation has been the original 1964 edition, published as *L'individu et sa genèse physico-biologique*, which I refer to as *L'individu*.[29] This edition compiles chapters on physical *and* vital individuation from both the first and second parts of the full thesis. The second part of Simondon's thesis (including further chapters on psychical and collective individuation) was not published in French until 2007, as *L'individuation psychique et collective*.[30]

It was John Protevi's book *Political Physics* that first led me to Simondon's critique of hylomorphism, which appears in the first chapter of *L'individu*. Protevi suggested a directly architectural reading of the schema by referring to Plato's discussion where the form-giver is rendered as an architect.[31] More compelling when I started reading Simondon's first chapter of *L'individu* 'Form and Matter', was the extended 50 page description of the processes involved in moulding a wet clay brick, used by him to replace the familiar hylomorphic conception of a brick shaped by the simple imposition of form on the passive matter of the clay, as if in an instant. Of course, it was serendipitous to find a material example drawn from the building site used to set up a critique of the form/matter schema, but what cemented my interest in this chapter was the remarkable similarity between Simondon's account of moulding clay, and some rather detailed process-based clauses for casting in reinforced concrete I had found in various 20th century specifications. Reading the two alongside each other revealed a shared sensibility and awareness between the found building documents and Simondon's description of the chains of processes involved in clay or concrete taking form. The specifiers who wrote the clauses were fully aware that concrete did not take form according to the simplistic terms of the form/matter schema, but nevertheless, 'above' the technical level of the specification, architectural discourse persisted in imagining concrete production in hylomorphic terms. This was evidence that in architecture too, the hylomorphic schema prevailed, despite the evidence of knowledge gained through experience in the workshop or on site, just as Simondon and Smith had argued with their respective disciplines.

In *L'individu* Simondon observes that philosophy has sometimes used technical operations as foundations for its universal schema, while at the same time overlooking that specificity. For example, Aristotle uses the casting of a bronze sphere as a paradigm to understand the being of all individuals in terms of form and matter, not just technical individuals, Simondon also starts with a technical operation – the forming of the wet clay brick – but uses the example to demonstrate the explanatory limitations of the hylomorphic schema. As the book develops the limitations become more obvious, when Simondon addresses other kinds of individuals, such as more complex physical systems, living organisms and finally (in *L'individuation psychique et collective*) human 'psychic collectives' or social groups. Hylomorphism is shown to be just as inadequate a schema for the moulding of the wet clay brick, because it does not account for the chains of energetic process that make the clay and mould ready for their encounter, or unfold during it. In Simondon's hands brick moulding is recast as a dynamic system through which individuation arises. It includes both the taking of form itself (also a chain of operations) as well as the preparatory processes that make ready a 'preindividual' state – in which there is potential for individuation not yet actualized. In Simondon's philosophy, individuation is common to living and non-living systems.

At the same time Simondon's premise in his explicitly technical work, *On the Mode of Existence of Technical Objects* (*Du Mode*), which finishes where *L'individu* started with the example of wet clay moulding, is that technics have not been assimilated as part of culture. A tool-based understanding of technics is inadequate, Simondon argues, because it suggests that our technologies merely mediate between humans and nature. They are of much greater significance than this, he claims. They are a new kind of nature arising at first from human thought. While the first iteration of a technical object tends to be a rather literal replication of the initial conceptual diagram (think of early telephones where each part – speaker, earpiece, dial – was a separate component) it 'evolves' through a series of adaptions to take advantage of possibilities suggested by the new object's physical manifestation (as did the telephone where the dial became the stand for a conjoined speaker/earpiece/handle, leading finally to the dial stand becoming redundant). In Simondon's thought these 'concretized' technical objects are a means by which our shared human collectivities are formed. Technics are to be understood, not feared or relegated to the level of mediating tool.

Thus at the start of this study, it was Simondon's general approach to technics and his analysis of brick moulding that first suggested his work could be relevant, especially to my discussion of the process-based clause, whose descriptions of mortars in the 18th and 19th centuries, and concrete casting in the 20th century resonate so closely with his re-description. But reading further, concurrent with my study of performance specifications, it became evident that his philosophy of individuation put process, and relationality at the centre of a framework that would gradually lead my thinking on building materials more generally.

Building Materials: An ontogenetic approach

For many of the commentators through whom Simondon's account of the wet clay brick taking form has become widely known within architectural theory, such as Deleuze and Guattari, Brian Massumi, and Lars Spuybroek, the focus is on what Simondon calls the 'dynamic operations' of form-taking. In the case of the wet clay brick, these are the energetic exchanges at the molecular scale and the exertions of force from clay molecule to clay molecule. Simondon understands form-taking as a self-organizing dynamic system that is contained and limited by the mould, not shaped by it as in the hylomorphic schema.

Hardly remarked upon, however, are another set of processes described by Simondon. They take place prior to the dynamic operations, and make both clay and the mould ready for their encounter. As Simondon points out, it is the specific plastic and colloidal properties of clay that enable it to take form in moulding. Wet sand poured into the brick mould would just be a heap of sand when it comes out of the mould. The potential for form-taking is not in sand, nor is it in clay when it is extracted from the ground. It must be prepared through another long, technically defined (and intentional) chain of processes, which include drying the clay, grinding it, filtering it, adding water, rolling it and kneading it. Simondon shows that clay – a material that seems at first to be a paradigmatic instance of matter – is in fact prepared and constructed towards this specific encounter with the mould. It is not simply a material to be formed or built with. It is a material that is itself built.

Simondon calls these processes 'preliminary operations' ('operations préalables'). Descriptions similar to his are also found in specifications for mixing up mortars and concretes, and preparing formwork prior to concrete casting. In these found texts, descriptions of processes are limited to materials that are prepared on site. For most building materials these preparations take place long before they arrive on site. Bricks are stacked ready for building, already fired and shaped. Where they are included in the specification, descriptions of preparatory processes remind us that materials must be made ready for building. Whether tree, metal or clay, they do not emerge fully formed out of the earth, ready to be transformed into built form. There are always processes involved that shape them towards their use in building. This is an argument the architect Peter Salter makes in *Intuition and Process*. A complex chain of operations such as steel-making is clearly a 'metamorphic process' but so too is the quarrying of natural stone. 'The first minute fissure in the material', writes Salter '[. . .] brings air to the fossil, [. . .] exposing a new plate for oxidization and weathering. The material is re-coded as it leaves the ground, gaining a new aspect and experience' acquiring 'the new material's range of possibilities'.[32]

That materials are prepared should not be a contentious claim. What is surprising is how often this fact is overlooked. Karl Marx, for example, writing about the homogeneous and plastic materials which money must be made out of if they are to embody 'abstract and therefore equal human labour', claimed that 'gold and silver possess these properties *by nature*'.[33] We know this is not true and that a vast range of laborious and detailed processes, from extraction through to refining and production, must prepare gold and silver for their capacity to be equivalent in every part and 'pure' for their use in money. Clearly this property is not given by nature, and indeed elsewhere Marx discusses the realities of extraction and production that brought about shifts in the relative values of silver and gold.[34] Perhaps in focusing his attention on the money commodity, it was all too easy to see the materials out of which it is made as *prior* to preparation and labour or in short, to confuse (raw) materials with the idea of matter.

The approach this book takes to building materials foregrounds these preliminary operations. I am more concerned with asking how materials are made ready for building and which factors guide those preparations, than in materials' qualities as substances and what can be made out of them. Out of the three forms of clause I examine, only the process-based clause describes these processes directly. The performance clause looks to the future behaviours of the material as definitive, not to the material's genesis, but nevertheless I show that vast sets of processes are needed to make it possible to describe a material in performance terms. In the case of performance-engineered materials we could even say that the intended use of a material is pre-programmed or becomes preparatory. Naming glass 'best seconds' or 'good thirds', also depends upon a set of preliminary procedures that test glass according to predetermined standards, managed in an institutional context. Even the names of materials sometimes reveal something of their genesis. For example, terms for timber used in 19th century specifications, such as 'Dantzic fir' or 'Christiana Deal', carry the trace of the ports where they were traded. Although Simondon describes a series of physical processes that make clay and mould ready, the study of the specification shows that 'extra-physical' processes such as testing should also be understood as preliminary operations. Indeed, the selection of one form of clause over another could itself be understood as preparatory. The argument presented in the later chapters of this book is that the use of one form of clause over another is more than a matter of simple choice. In fact, the use of one of these forms of clause signals that the material has been subject to a characteristic constellation of preparatory operations particular to the type of clause.

Admittedly, preliminary operations are little more than a passing concern for Simondon. The 'Form and Matter' chapter is the only place they are mentioned in *L'individu*. Although Simondon goes on to describe other physical and manmade systems in his next chapter 'Form and Energy', including a pendulum or his famous example of a crystal growing in its solution, he shows no interest in their preparations. Describing these examples, he focuses only on the unfolding of dynamic operations of the individuating system. He is explicit however, that his attention to process is what defines his philosophical approach. In a dense appendix entitled 'Allagmatic' added at the end of *L'individuation à la lumière des notions de forme et d'information* (2006), Simondon states that his aim is no less than the substitution of the theories of structure with which philosophy has long been concerned, with a new theory of process. A problem with the hylomorphic schema, as he explains in the Form and Matter chapter, is that they are inadequate to explain the wet clay brick, since form and matter are themselves already individuals. An account is needed of the process of becoming individual that does not rely (illogically) on already individuated individuals. Thus, Simondon shifts attention from the individual to how the individual comes into being; from being to the genesis of being; from the study of being (ontology) to ontogenesis. In *Du Mode* it is again how technical objects come into being, or evolve that concerns him. And in doing so it becomes apparent that individuation is common to both physical and living beings. Similarly, this book does not attempt to make any claims for the vitality of matter, nor is it concerned with a distinction between living and non-living materials. Rather it takes an ontogenetic approach to building materials in the sense that it is concerned with how they come into being and with understanding that materials are themselves built.

'Veritable Relations': Building materials as systems

Much better known than Simondon's is Martin Heidegger's critique of the hylomorphic schema in his essay 'The Origin of the Work of Art'. Where Simondon and Smith situate the origin of the schema in the workshop, Heidegger sees it as deriving from the relationship of the artist to the artwork, for whom matter is merely, 'the substrate and field for the artist's formative action'.[35] The artwork however, according to Heidegger, is exceptional. Most things, such as jugs, axes and shoes, are made to be used. When something, 'is made as a piece of equipment' it is made '*for something . . .*' and in these instances, 'matter and form are in no case original determinations of the thingness of the mere thing.[36] Matter and form on their own are insufficient explanations for the 'thingness' of equipment. What a piece of equipment is intended to be used for predetermines its genesis, even if finally it is used in another way. Material choices for equipment, says Heidegger, are therefore 'grounded in such usefulness' and he gives some examples: 'impermeable for the jug, sufficiently hard for an ax, firm yet flexible for shoes'.[37] Here the material to be used is not named. It is irrelevant whether the jug is to be glass or china, or how it is to be made. All that matters is that it will be watertight. We do not need to know if the shoes will be made out of leather or plastic; what is determining is that the material should both protect the foot and allow for movement. Although Heidegger uses the term 'prescription' for this preparation for use that characterizes equipment, we can say they are simple instances of performance specification where a material is described for what it is to do. If Simondon's process-oriented critique of hylomorphism has given us a philosophical analogue for the process-based specification, then

Heidegger's equipmental philosophy of technology seems to provide a remarkable parallel for performance specification (explored in Chapter 5).

It is rare to find performance specifications in the documents of building practice that read as quaintly as Heidegger's phrases. One instance appears in a specification for a gothic villa for Mr Usborne (1879), where the drawing room timber floor is specified only in so far as it may be 'partially traversed for dancing'.[38] No details are given that explain how this property is to be achieved. The carpenter must just ensure that it will be sufficiently springy for the occasional dance. Contemporary performance specifications read very differently. They adopt the quantitative language of materials science, a discipline that emerged out of the study of polymers and was only established in the 1950s and 1960s. In these specifications material properties such as acoustic insulation, heat or UV light transmission, strength in compression, traffic noise resistance and so on, are defined using scientific units of measurement – Megapascals (Mpa), Relative Humidity (RH), Watts per Metre squared Kelvins or u-value (W/M^{2}K) – and stated in quantities, grades and percentages. Nevertheless, it is how the material is to be used that is decisive for these performance specifications too, whether to admit light or retain warmth and so on. As Heidegger argues for equipment, in performance specification the material is 'grounded in such usefulness' as part of its preparations.

With the advent of performance-engineered materials, the prescription of use has become more literal. Through the sandwiching of microfilms and interlayers, a material like glass can now be designed to achieve very precisely defined behaviours. Of course this involves testing the material's properties to measure its capacity to withstand load, to absorb sound or transmit heat and so on. But in addition, performance-engineered materials require that 'goals' (to use Bruno Latour's term for a programme of action that could be intended by both human and non-human actors) are physically embedded in the material's structure. This is significant for the arguments presented here in two ways. First, a performance-engineered material like glass has undergone a process of 'pre-inscription' (as Latour names the embedding of goals into technical artefacts such as door closers or speed bumps) that designed and engineered the material to achieve specific levels of emissivity or reflectivity. In this sense, although also a material, it is like any other technical object or artefact that has been designed and pre-inscribed. Second, following Simondon's account of clay being made ready for the act of moulding, we begin to see that even more traditional building materials are already built and prepared *for* something. This challenges the reductive notion of building materials as matter or substrate for receiving form. Building materials do not simply exist prior to their being shaped. Rather, like any other technical object in Simondon's definition, they too are new additions to the natural world, thought and made by man. A central premise of this book is to consider building materials as technical objects.

In Heidegger's terms however, the recognition that building materials are prepared 'for something', as clay is 'for moulding'; or glass 'for emissivity' and so on, entails that they are equipmental. Simondon suggests something similar in his proposition that a 'form intention' guides the preparations of the clay, but more generally his position in *Du mode d'existence des objets techniques* (1989), his major work on the genesis and being of technical objects, is explicitly in opposition to Heidegger's. To think of technical objects only in terms of their use is to miss significant differences between them. A clock with weights and an electric clock have a common use, but the first is more like a windlass, Simondon points out, while 'an electric clock is analogous to a house-bell or buzzer'.[39] If we are to have more integrated, less alienated relationships with the technical objects and systems we live with, we should, he argues, take into account the technical object's (concrete) physical structure and operations. We should understand its functions with respect to other technical objects and the humans that work with them. And moreover, we should examine the genesis of a technical object, seeing it, 'as the end product of an evolution' and not as something which can 'be considered a *mere utensil*' (and here Simondon refers directly to Heidegger's term 'utensil').[40] We might be more alert to the opportunities technical objects offer for further adaptation and evolution. The aim of *Du Mode* is to set out an ontology of technical objects that avoids defining them through use.

These contradictory philosophical positions suggest that the process-based and performance clause are premised on very different – even mutually exclusive – concepts of materials; one process-oriented and one equipmental. Despite this philosophical incompatibility, it is common to find both forms of clause side by side in specifications. Indeed it could be possible to describe the same material in either form of clause (although as we will see in Chapter 5 it is not so straightforward). Could we simply conclude that the variety of forms of clause is evidence of a variety of concepts of material? Rather than stop at the conceptual incompatibility between process-based and performance specification, I draw on another aspect of Simondon's philosophy to suggest that the variety of forms of clause points to understanding materials in terms of the variety of kinds of relations they engender, between say clay and a mould, or glass and its environment. Simondon puts real communication or mediation – or 'veritable relations' (as he refers to them in the Form and Matter chapter) – at the centre of his account of individuation.

It was researching the performance clause, and the preparatory methods of testing materials that led me to pursue Simondon's arguments about relations. Say for example an architect is designing glass curtain walling next to a park, and she wants to ensure the glass does not break when a football kicked with force hits it. In order to determine the strength of the glass to specify, the force of the ball must also be quantified. The real practices of quantification (testing on both sides) establish equivalence between the ball kicked by a child on a Sunday morning and the glass wall of the empty office building. Although it is only the property of the material to be used that is given in the clause, performance specification always also implies the material's relationship to another factor; whether this is the wider building structure (as in the case of the building's self load, or constructional performance); the physical environment (wind, rain, sun, pollution and so on); the social environment (footballs, security threat, sound pollution etc.) or inhabitation (temperature, light, ventilation and other aspects of comfort). Performance specification, including the many practical and technical process that make it possible, is a process that establishes compatibility between each of these factors among others, and the materials we use to build.

For Simondon, individuation is more than the unfolding of dynamic operations. What occurs during individuation is a mediation between two scales of magnitude that were previously incompatible. During the form-taking of the brick for example, while the expanding clay is restrained by the mould, clay and mould are in communication with each other. Once form-taking is complete their communication also comes to an end. The mould and the now-formed clay brick are separate entities, with no potential for further communication (you would have to knead the clay again to start the process over). Some kind of mediation between two scales or states, is common to all the systems of individuation Simondon discusses; the outer layer of the solid crystal becomes compatible with the liquid solution while the crystal is in formation; photosynthesis is the capacity of the living being 'to bring different orders or magnitude into relation with one another: that of the cosmic level (as in the luminous energy of the sun, for example) with that of the intermolecular level';[41] for a non-alienated worker it is the possibility to have an iterative, feedback relation with the machine they operate. And in each case communication comes to an end when individuation ceases; when the crystal solution is used up; or the worker is only monitoring the operation of the machine. For Simondon, as Alberto Toscano puts it, 'to relate – orders of magnitude, differences of potential, and so on – simply *is* to individuate'.[42]

In addition to individuation and its cessation, Simondon identifies a third state prior to individuation in which there is already potential for mediation, as yet unresolved. For example, when clay is ready for forming, but has not yet been rammed into the mould, or the seed crystal is yet to be dropped into the crystal solution. He refers to this as the preindividual state, or the condition of 'metastablity', neither stable nor unstable, like the snowy slope just before a sudden sound or a skier (acting analogously to the crystal seed) sets an avalanche in motion. In the preindividual state the potential for compatibility between disparate orders must already be there even if as yet unresolved. For example, in the case of the crystal seed it must already share some structural identity at the molecular level with the crystal solution that allows crystal formation to begin. A shard of some other substance could not catalyse growth, and would just sink to the bottom of the jar.

The mediations Simondon describes are real – or concrete, not abstract. They are 'veritable relations'. In the case of the clay brick, real energy exchanges at the molecular level are transmitted through the mass of clay as its surface meets the limiting form of the mould. That clay and mould can come into communication in this way is given through the preparations that rendered the possibility of their compatibility. In this case, preliminary operations 'created the conditions' as Simondon puts it, for communication to arise, between the clay and the mould most immediately, but also less directly between the clay and the geometric idea of a rectangle.

Chapter 6 (Systems of Material) follows closely Simondon's account of individuation as relation and his notion of the 'complete system' of individuation which includes within it the preindividual state, to propose that what we catch sight of through the radically distinct forms of clause found in architectural specifications are sets of processes that construct materials' capacities to mediate relations in very different ways. In the case of concrete casting, the process-based clause is concerned with how liquid concrete and formwork can come into relation through a careful sequence of preparations and detailed adjustments. The mediation is not just between shuttering and poured concrete; in the architectural context, the geometric idea of the designer, probably drawn, comes into communication with a slurry of pebbles, cement and water. These chains of processes are better understood in terms of a material system constituted by the real formation of relations, than in the reductive terms of form and matter. Similarly the complex chains of operation involved in testing footballs and glass are what enables a material to be selected or engineered for its structural performance. Many processes are involved in the chains of tests, calculations and scientific analysis that bring football and glass into communication are 'extra-physical', a term I prefer to 'immaterial' for operations that do not involve physical substances and manipulation, since strictly speaking, all these processes are material. Nevertheless the extra-physical processes that make performance specification possible forge relations no less real than the physical mediations in concrete casting.

To consider building materials as systems involves including in our conceptions of them the contexts in which they are produced and the ways they are to be mobilized. A sheet of regular glass in the window of a garden shed might look only imperceptibly different to glass that has been designed, tested and certified to withstand a certain bomb blast, but the regime in which it was produced limits the ways it can be used as much as its material properties. A straw bale might be able to withstand structural loads or provide insulation in a wall, but if it has not been standardized and performance-certified other materials will have to take on these roles to gain building control approval. Even if it is actually carrying the load, another set of operations are required for that relationship to be formalized.

Materials, at least as they are used in building, are not only defined by their physical properties and characteristics. We should be wary of the maxim 'truth to materials' that is as common to designers working with the emergent properties of living materials today, as it was to modernist architects such as Louis Kahn.[43] We should no more understand materials as mere substances distinguished by their physical characteristics, than in the limited terms of homogeneous matter. Technical, contractual and bureaucratic procedures are integral to materials in the wider sense of them as systems. They are 'informed matter' to use a term coined by Andrew Barry (Chapter 5). These factors determine how building materials can be known, specified and used, whether or not we choose to engage with them in our architectural theories and design practices. As Emily Thompson demonstrated so conclusively in her ground-breaking book on acoustic control in the US, *The Soundscape of Modernity*, building materials were already being designed and manufactured to achieve social goals as early as the beginning of the 20th century.[44] With the advent of performance-engineered materials, an increasing range of social norms from security to self-cleaning are embedded and reproduced in the fabric of the built environment. Surely architects should be paying more attention to these changes rather than accepting a status quo in which the selection of materials is handed over to contractors and specialists.

On the Transductive Method

One of the more pleasing discoveries of the survey of historical specifications undertaken for this book has been how rarely it seems possible to apply just one type of clause throughout a single document. Specifications exhibit a marvellous persistence of variety in their forms of clause, just as on the building site we find archaic building processes next to the most technically advanced. Even in the performance-oriented specifications developed by Schumann Smith in the last couple of decades, some materials and some qualities resist 'open specification'. For example, they invent a curious type of specification – 'indicative material' – to specify quarried stone without actually naming it; and revert to old-fashioned process-based specification to describe how glass is to be kept transparent and clean. No singular regime of production or description seems able to tame the variety of building materials. They resist any overarching universal notion of matter.

Attention to specificity and difference is a matter of principle in Simondon's work. *L'individu* is full of long descriptions of different systems of individuation;

from pendulums and dividing single cell organisms to corals and batteries. The detailed analyses of technical systems in *Du Mode* are illustrated with tiny grainy photographs of engine motors and their parts, motorbikes, triodes, telephones, windmills and turbines, covered with his scratchy handwritten annotations. Simondon's lectures in *L'invention dans les Techniques* are accompanied by copious freehand diagrams of transformers, alternators, brick vaults and gothic buttresses, Neolithic tombs, bridges and lamps. Needless to say, he attends to these cases not as objects or substances, but to their processes of individuation, whether in the minutiae of changing states as in the formation of the crystal or in the long arc of technical evolution as objects inserted into the world that in turn suggest new iterations.

In part, Simondon's recourse to describing one individuating system after another is a kind of demonstrative repetition. It makes evident that modes of individuation occur across non-living systems, living systems and in the formation of psychosocial collectives, and demonstrates the purchase of his approach. Indeed Jean-Hughes Barthélémy proposes this 'encyclopedism' that 'thinks the genesis of each thing' is a central ambition of Simondon's.[45] Barthélémy explains that its humanist inclusivity enables Simondon to develop an alternative attitude to technology than alienation. But it can also be understood as a method developed by Simondon as a response to the particular problematic of attempting to look at individuation rather than at the individual. How is one to move away from a model of research that considers only the already individuated individual, or even from one that deduces the processes of individuation from that individual? Neither induction, nor deduction are appropriate to this task, says Simondon, 'we will have to employ both a new method and a new notion' – 'transduction'.

Simondon's primary use of the term 'transduction' is in his accounts of physical systems of individuation where a change of state occurs. The latent energies in a crystal solution become structure in the formation of the crystal. Structure becomes energy in the case of an electronic signal. As Adrian Mackenzie explains, Simondon draws on ideas that emerged in the molecular biology of the 1950s[46] and from electronic engineering;

> In electrical and electronic engineering, transducers convert one form of energy into another. A microphone transduces speech into electrical currents. For the process of transduction to occur there must be some disparity, discontinuity or mismatch within a domain, two different forms or potentials whose disparity can be modulated. Transduction is a process whereby a disparity or difference is topologically and temporally restructured across some interface. It mediates different organizations of energy. The membranes of the microphone move in a magnetic field. A microphone couples soundwaves and electrical currents.[47]

The transductive process also has a more general meaning for Simondon. It is any 'individuation in progress'.[48] As he explains in L'individu it can equally be applied to thought. Simondon asks us to imagine a mode of thought that unfolds in an analogous way to the crystal growing in its mother water, layer by layer, with each layer provides the structuring basis for the next. The crystal's growth is 'a progressive modification taking place in tandem with it'.[49] Structure and individuation – the energetic process of communication between the previous layer and the solution – are one and the same. Simondon develops this as an image of thought unfolding as two domains come into relation with each other.

> Transduction does not seek elsewhere a principle to resolve the problem at hand; rather, it derives the resolving structure from the tensions themselves within the domain just as the supersaturated solution is crystallised due to its own potentials and the nature of the chemicals it contains, and not through the help of some foreign body . . . Transduction represents a discovery of dimensions that are made to communicate by the system for each of the terms such that the total reality of each of the areas' terms can find a place in the newly discovered structures without loss or reduction.[50]

Transductive thought arises not through applying pre-given abstract concepts to a situation, but through the process of thought emerging in tandem with the system it considers.

> Clearly transduction cannot be presented as a logical procedure terminating in a conclusive proof . . . I see it as a mental procedure, or better the course taken by the mind on its journey of discovery. This course would be *to follow the being from the moment of its genesis, to see the genesis* of the thought through to its completion at the same time as the genesis of the object reaches its completion.[51]

In transductive thought problematics and its conclusions are at once the process of putting them together and following them through. Transductive thought must therefore need and retain the concrete systems it studies and works through. We might also recall Smith's notion of a 'funeous' approach to materials – that does not reduce them to abstractions in order to understand them through other concepts, but retains their specificities and histories. Moreover, transductive thought avoids being hylomorphic in as much as it does not impose a conceptual structure from elsewhere, but finds its problems and its resolutions from within its system. In Simondon's sense transductive thought is also inventive.

In the area of knowledge [transduction] maps out the actual course that invention follows, which is neither inductive nor deductive but rather transductive, meaning that it corresponds to a discovery of the dimensions according to which a problematic can be defined.[52]

Simondon's notion of transductive thought draws on the use of the term by the child development psychologist Jean Piaget, who argued that children's thought neither applies a general rule (deduction) nor derives one from a number of instances (induction) to account for a situation, but 'moves from particular to particular by means of a reasoning process which never bears the character of logical necessity'.[53] Transductive thought is not just intellectual; it is also intuitive. It 'is not only a path taken by the mind, it is also an intuition, since it allows a structure to appear in a domain of problematic yielding a solution to the problems at hand'.[54]

The method pursued in this book of relating the forms of clause in the architectural specification to the question of concepts of materials and to Simondon's philosophy of individuation can also be seen as transductive. It is not supported by a particular research method or the application of a pre-existing theory of materials, nor is it a logical process.[55] Although it might be described in such terms – that a technical document should not be precluded from theoretical study, that the descriptions it contains can be argued to be conceptualizations, that the industrial context must be made primary, that the specification is complementary to the architectural drawing and so on – this choice to study the specification has not derived its validity from some other set of principles. It is not, 'a logical procedure terminating in a conclusive proof'. Rather the act of making the specification a vehicle through which to think about architectural materials brings a set of questions into being which are particular to this 'system' or process of thought. It would be impossible to avoid, say the context of the building industry and changes in production, or the question of the variety of forms of clause. The process, in as much as it sets up a system for the research, brings a set of questions into being which did not exist prior to the attempt to follow the process through.

As such, the book is structured, in so far as it is possible, to follow, 'the course taken by the mind on its journey of discovery'. I do not mean that this is a chronological presentation of the research as I undertook it, rather, that instead of foregrounding an account of Simondon's philosophy and then applying it to the forms of clause in the specification, his work is introduced gradually through the chapters as aspects of it relate to the discussion of forms of clause in the specification. Chapter 2 'Specifying Building Materials' introduces the changing formats of the specification in the UK through time, in order to set out the contexts in which forms of clause emerge and are used. Chapter 3 examines

'Naming', the most prevalent technique for specifying building materials, which appears to confirm the widely held notion that particular materials are simply diverse manifestations of a 'universal concept' of matter. Simondon's philosophy appears only as it becomes active, or in communication with the particular research domain of this book. This starts in Chapter 4 'Process' with the analogous descriptions of the wet clay brick (Simondon) and casting concrete (specifications), and takes more of a lead in Chapter 6 'Systems of Material'. In light of the centrality of relation to Simondon's account of individuation, the variety of forms of clause is understood as evidence that building materials might better be understood as individuating systems. Finally, Chapter 7 'Going Into the Mould' turns to Simondon's call for more inventive, less alienated relations with the technical objects and systems we live with and amid, including building materials.

Some readers may come to this book for its study of the specification, in which case you can follow Chapters 2, 3 and the relevant sections of 4, 5 and 6. Others may be more interested in the account of Simondon's work in which case you could start with the last section of Chapter 3 where Aristotelian hylomorphism is introduced via a table for indexing technical literatures on building materials in architect's libraries, and go on to read the relevant sections of Chapters 5, 6 and 7. To pursue the ontogenetic theory of building materials offered here, you can simply go along with the book's structure and '*follow the being from the moment of its genesis, to see the genesis* of the thought through to its completion at the same time as the genesis of the object reaches its completion'.

A great potential of transductive research is that the domains studied contribute to the genesis of thought in such a way that the differences between them – their concrete realities – 'can find a place in the newly discovered structures without loss or reduction'.[56] As Simondon claims, 'nothing proves in advance that there is only one possible way for the being to be individuated'. If we were to find, he continues, that 'many types of individuation existed', then, 'similarly there ought to be many types of logic, each corresponding to a definite type of individuation'.[57] In this book I show it is not sufficient to explain the variety of forms of clause in the specification – whether the classificatory practices of naming, or the method-oriented practices of process-based description, or the quantitative techniques of materials science – in terms of the application of epistemological regimes from outside. By considering them as different conceptualizations of materials that arise through technical and industrial activity, and exist as a plurality rather than as any one overarching conceptualization, their variety leads, via a reading of Simondon, to a theory of building materials as individuating systems and the possibility that we might need more than one logic to account for them. In such a theory, how we think about materials cannot be extracted from the conditions in which they are produced. To recognize this variety of concepts is at once to insist on the centrality of the material's preparations and their changing conditions but also to suggest that each system of material offers very different possibilities and starting points for designing with materials in ways which might recognize the range of relationships that are mediated beyond simply giving presence to a formal idea or conforming to 'visual intent'.

As an alternative to the imposition of frameworks from other disciplines such as philosophy or pure science, the transductive method has potential as a model for thinking about an applied discipline such as architecture that attempts to retain 'all the concrete' and 'comprehends all of the concrete',[58] rather than losing the concrete in the move towards abstractions. It could enable the discovery of concepts of materials that are particular to applied disciplines such as engineering, chemistry and architecture, at once retaining disciplinary and practice-based specificity and recognizing the potential for such practices to produce their own concepts. But of most benefit, as Simondon argues at the end of *Du Mode*, may be its potential to 'go into the mould' and open up the black box of our relations with technical objects and systems, in order that we are better equipped to understand them as they exist now, and to invent with them other less alienated possibilities for working with and existing alongside them.

2

Specifying Building Materials

As John Gelder's research has shown, the use of text-based lists of the materials to be used in building work – whether accounts, bills of quantities, specifications – both predates the use of the architectural drawing and has been far more prevalent. However, these documents are much less likely to have been retained. Just as Sigfried Giedion complains in his foreword to *Mechanization Takes Command*, the discarding of technical literatures has been commonplace, in his case compromising his efforts to write an anonymous history of industrial methods. These 'acts of destruction' he writes, were the 'murder of history'.[1] Even where specifications have been kept and are available for study, historians of architecture and construction have tended to read them as conduits to learning more about a building's construction, or – more rarely – to better understand the business of building, management and regulation. Outside guides to best practice, it is very rare indeed that attention is paid to these documents as artefacts in their own right.

This chapter studies changes in the contents and organization of the specification over time. While we can expect changes in what is being described due to technological developments in building, which give us insights into construction practices, this chapter will identify more general kinds of inclusion and exclusion at the level of the detailed clauses of the document, or 'statements' (to use Foucault's term), and at the level of the document's overall emphasis, or the context in which the statements acquire meaning. The chapter demonstrates that much more significant – even conceptual – changes can be detected in the form of the specification and its clauses, than the simple fact that old technologies and techniques disappear from the descriptions and new ones emerge. It is no surprise to find, for example, that in 2009 a work section entitled 'piped supply systems' replaces instructions to 'well-sinkers' in the 1898, or that 'communications / security / control systems' replace 'bell-hangers'.[2] Or similarly, that in the early 19th century references to concrete appear only in the 'foundations' division, whereas the material is now ubiquitous throughout the building process and is likely now to appear in many sections – from groundworks to cladding and surface finishes. The work carried out by the various trades changes too. Instructions to plumbers are relocated to another part of the document when lead-working is no longer their main trade, and sections for asphalters become more prevalent with the introduction of the modernist flat roof.[3] But we can also detect more radical changes in the scope of the specification, in what is deemed relevant for inclusion in one period, and entirely omitted in another, or in the manner in which a material is described, that alter the emphasis of the document and its concerns.

Thus we should be circumspect in assuming that the job of the specification is simply, as specification consultant Andrew Knightly Brown explains, to be 'specific'. It turns out that the nature of that specificity is by no means constant or universal across documents:

> One of my first tasks with any new client is to remind them that the word 'specification' includes the word 'specific'. Look up synonyms for the word 'specific' and the words 'precise' and 'explicit' are often listed; look up a definition for 'explicit' and you'll find the phrase, 'fully and clearly expressed, leaving nothing implied'. 'Leaving nothing implied' – what a great phrase to describe the essence of a specification.[4]

That the nature of *what* is 'fully and clearly expressed' varies, and in how it does so is demonstrated by the following examples. In the UK a 19th century specification was usually divided trade by trade according to the order works were to take place on site – 'Excavator', 'Well-Sinker', 'Concretor', 'Bricklayer' etc. (see Fig. 2.3). Within each 'trade-division' clauses describing individual parts of the building work are rich with details of how it was to be carried out on site, often given in lengthy prose. In contrast, the 21st century document produced by Schumann Smith – a specialist company who specialized in the production of litigation-proof specifications for large-scale building projects that might otherwise risk incurring significant claims – is organized in work sections with titles that emphasize the products, components and systems of building rather than the work of the trades – Section C: 'Demolition/ Alteration/ Renovation', Section D: 'Groundworks', Section E: 'In situ concrete/ Large precast concrete', Section F: 'Masonry' etc.[5] More strikingly the clauses of this document may not name the materials to be used and tend to avoid specifying any building techniques. Performance criteria – given in the quantitative language and units of materials science – comprise the main body of many clauses. In these documents processes of building have been intentionally omitted to hand over responsibility to the contractor to find methods and details that will comply with the performance criteria. Clearly the nature of information available to the historian reading either of these specifications is very different. The trades-based document is a useful resource for understanding the processes of building, while they are almost invisible in the Schumann Smith document. Indeed it has been carefully constructed and worded to avoid any specificity about methods of building.

The changes in the arrangement of the written documents, the kinds of detail specified and the language used cannot be explained simply by new developments in construction techniques. In what manner and to what extent the specification is specific, is also a product of the contractual context in which it performs. The clauses of the specification and the structure through which they are organized should be understood in terms of 'statements' as Foucault outlines them in *The Archaeology of Knowledge*:

> In the seventeenth and eighteenth centuries Natural History was not simply a form of knowledge that gave a new definition to concepts like 'genus' or 'character', and which introduced new concepts like that of 'natural classification' or 'mammal'; above all, it was a set of rules for arranging statements in series, an obligatory set of schemata of dependence, of order, and of succession, in which the recurrent elements that may have values as concepts were distributed.[6]

They are linguistic performances whose coherence is not given through logic or grammar, but are linked to each other and require for their operation, as Foucault puts it, an 'associated field' or 'domain of coexistence for other statements' and a 'materiality (which is not only the substance or support of the articulation, but a status, rules of transcription, possibilities of use and re-use)'. The writing of the specification and the form it takes depends on more than the object under description – the building and its various parts – or the preferences of the writer. Dominant regimes of knowledge, whether the craft-oriented descriptions in trades-based 19th century documents or the units and quantifications of materials science in performance-based 21st century documents, are not sufficient to account for differences between documents. Rather, the meaning and force of the specification's statements and their arrangements are tied to the contemporaneous institutional norms and practices of building. As Foucault argues, 'The statement cannot be identified with a fragment of matter; but its identity varies with a complex set of material institutions'.[7]

In the case of building the set of material institutions through which particular practices of specification arise is indeed complex. In addition to the material changes in techniques and materials, and the scale of what is built, there are changes in the ways contracts are let out, in the division of labour and roles between guilds, artisans, building workers, contractors and architects and the legal obligations of each of these parties. How building is financed and the degree to which contractual sums are agreed in advance varies, as well as the extent of building legislation, regulation and standards. The trades-based specification of the 19th century, for example, developed with the emergence of 'contracting in gross' where a general contractor gave a price in advance of building, based on the details of the specification and the full set of drawings. The contractor's profit depended on being able to let out work to individual trades and procure building materials at a cheaper price than the amount he tendered for. The specification was an important tool to ensure quality of work and prevent the contractor from cutting corners. The architect who wrote the specification had to spell out the proper methods of work in full for each of the subcontracted trades. Today the building context has changed for large-scale projects in two important ways. The Design and Build contract has been introduced to allow contractors to secure profit by producing their own detailed design that best suits their methods of working and access to materials and components.[8] Moreover, the pursuit of claims by contractors has escalated, to the extent that claims can be made because of inconsistencies within the documentation (even if they have no implications for the building as it is realized). It is not just profit that motivates a different kind of (non-specific) specificity in Schumann Smith's specifications, but also the need to avoid litigation.

By making a broad survey of the changing arrangements of specifications from the 18th century until today, this chapter shows the extent to which the form of description is embedded in the practices of building and its material institutions. Present-day technical and practical literatures do recognize that different contractual arrangements require different practices of specification, but this point is rarely made with respect to historical developments. Davis, for example, notes more than once, an increase in detail and 'specificity' in specifications and other contractual documents written since the emergence of contracting in gross.[9] In general, there has been a tendency for specifications to become ever lengthier but this is by no means always the case. In 1769, for example the specifications for the government-funded building of Newgate Gaol and the Old Bailey Sessions House in central London, comprised several pamphlets, one for each trade, with the most detailed for the Carpenter already 8 pages long.[10] A year later, the specification for a new house and bakery for the gingerbread maker John Steinmetz, also in central London, follows the traditional format of a signed and sealed indenture written on a single sheet of parchment.[11] In the 1930s, a specification for a 'smoked sausage manufactory' is no more than a list of materials typed hastily on a few sheets of foolscap that identifies for the contractor which of the most expedient construction techniques to use for the quick assembly of a factory shed.[12] But in the same decade, the specification for an architect's own house using the then-new techniques of in-situ concrete casting runs to 40 pages. In each of these cases the method of procurement and the contractual organization that is more decisive for the degree of precision than the date of building. Moreover, in each of these cases, the kind of detail considered relevant also varies.

Individual building projects will always provide exceptions to generalizations and at any given time many forms of contract are in use and therefore, necessarily, more than one form of specification. Nevertheless this chapter identifies four key phases in the arrangement and scope of specifications since the 18th century. The first type will be exemplified by a 1734 indenture for a town house at St James Square and describes the building as an object, much as a drawn section would do, and relates to the practice of paying a master builder by measure after the work has been realized. The clauses of this early kind of 'dimensions-led' specification give no detail about methods of work and rely almost entirely on 'naming' materials that will be discussed in Chapter 3. The second type – the trades-based specification – is usually identified with the emergence of contracting in gross in the 1830s but in fact appears in the later decades of the 18th century, due to new practices of fixing contract sums in advance of building. To specify materials it makes use of naming, but also introduces detailed descriptions of methods of work or 'process-based' clauses that will be the subject of Chapter 4. These documents are broadly process-oriented, in contrast to those of the early 18th century, and their use flourishes until the 1960s, when the first time in the UK a concerted effort was made by the RIBA to produce a standard 'skeleton specification' for use throughout the profession. The aim of the group who developed the new National Building Specification (NBS) finally launched in 1973, was no less than to 'state requirements in terms of ends rather than means'.[13] In other words, descriptions of the processes of building or 'means' were to be replaced with descriptions of the end result. The structure of the NBS reflects this 'ends-oriented' description, but in practice, their skeleton documents have tended to use the full range of forms of clause, including process-based and performance specification. The newer type of specification is exemplified here by the 'Specification for the H. . .y Building' produced for a large office building in the City of London in 2006 by specification consultants Schumann Smith (who have since split). Their efforts to produce documents that avoid 'prescriptive' or 'closed' specification such as naming materials or process-based specification characterize the fourth type. To achieve 'open specification' for design and build contracts within the intensely litigious context of large-scale building today, Schumann Smith make considerable use of performance specification which will be explored in detail in Chapter 5 and a further category – 'visual intent' – that will be discussed in Chapter 7.

Of significance for the arguments of this book is the fact that these four types of specification – dimensions-led; trades-based; ends-oriented; open – each arise out of specific institutional frameworks of building practice that determine the arrangement and meaning of their statements. At the same time the descriptions they yield – particularly at the detailed level of their clauses – are *conceptually* different. Buildings, and the materials they are made from, may be described variously as dimensioned objects, as processes, as 'ends', through the performance criteria or behaviours they are to meet, or through 'visual intent'. To return to Foucault, the changing nature of the specification at the general level and the forms of those clauses that I have begun to identify at the more detailed level can be considered in terms of his notion of the concept. Just as, for Foucault, the mammal is a new concept in as much as it does not refer to the discovery of a new animal but to a new way of thinking about animals, a change in the form of clause need not relate to the production of a new kind of material, but rather constitutes it anew within the transforming contexts of building production, financing and administration. As Foucault makes clear in relation to natural history, if we are to understand these new concepts we also need to look at the kinds of rules determining the ordering of statements in relation to each other and at their associated field. 'The coordinates and the material status are part of its intrinsic characteristics', he writes.[14] Concepts and statements, in his terms, cannot be separated out from the ways they are organized and structured in relation to each other, nor should they be abstracted from the conditions of their emergence or their effects:

> [Statements] are invested in techniques that put them into operation, in practices that derive from them, in the social relations that they form, or, through those relations, modify.[15]

The changes in the specification and the forms of clause I identify are not always limited to one period of time (and indeed one of the questions this study raises, contra Foucault, is why it is that some descriptions of materials in the specification do *not* change, even when practices and knowledge formations undergo radical transformations). But the documents they appear in show enormous variation, at the level of their material status, in terms of their organization and in the ways they are put into operation.

From Object to Process: 18th and 19th century specifications and the shift to division by trade

The earliest specification held in the RIBA archives – The Articles for Agreement for a town house for Sir William Heathcote (1734–6) in St James Square, London – is effectively a drawing of the building in words (see Fig. 2.1). It appears rather handsome, written on heavy parchment, sealed with wax, and signed by Heathcote, his builder the master carpenter Benjamin

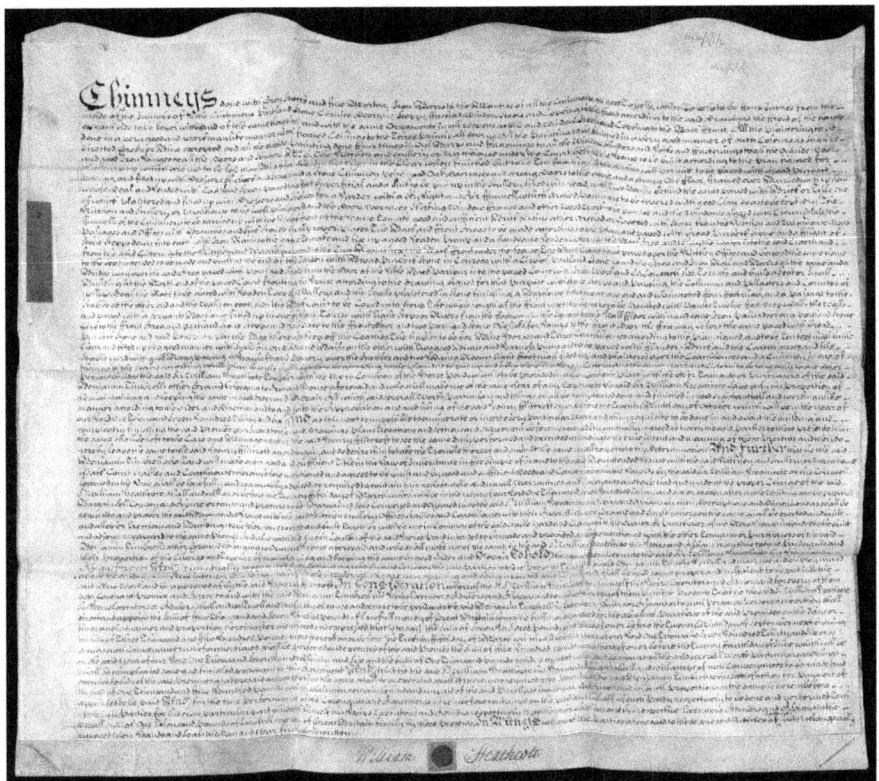

FIGURE 2.1 Articles of Agreement for a town house for Sir William Heathcote at St James Square (1734–6), RIBA Archives, HeW/1/1/2. Source: Courtesy of the Royal Institute of British Architects Library Drawings and Archives Collection.

Timbrell and the architect Henry Flitcroft, who had himself been a carpenter. The notary's script is ornate and continuous, leaving no spaces between words, indicating that the text was never intended to be read. Its form is that of a medieval indenture or contract, and would have been written out in duplicate before being cut with a wavy line keying both halves and forever signifying their original union. A sentence in the document prescribes that one half would be kept by the architect, the other presumably by the carpenter.

Both Davis and John Summerson[16] include this document in their accounts of Georgian building in London, and note the rather general nature of the indenture's description of the works. Building materials are only sometimes named, and there is no mention of methods or quality to be achieved, as in the following clause:

> No.3 Plan of the Chamber floor. The same to be thirteen foot six inches high in the clear when finished the teaked flooring framed of sound and good fir timber Beams fourteen inches by twelve the ceiling joysts a little below the Girder to preserve the Ceilings from cracking the upper joyst not less then neine inches by two and a half to be floored with clean deal dowelled except the landing of the two staircases[17]

As in this extract, most of the information given in the document concerns the dimensions of the timber structure on each floor. Moreover, the description follows the structure of the building. It is given floor by floor in the following order:

> No.1 A plan of the lower storey
> No.2 A plan of the hall and parlour floor
> No.3 A plan of the chamber floor
> No.4 A plan of the attick storey
> No.5 A plan of the garrets or upper storey
> No.6 The kitchen and scullery or washhouse under the court behind the house.[18]

The organization of this specification is structured much as a drawn section of the building would be – from the lowest point in the house, the cellar, to the highest, the garret, with the outbuildings last. The building is described as an object, without the references to quality, workmanship or methods that will characterize the lengthy specifications of the 19th and 20th centuries. Both the arrangement of the document, and the manner through which the building is understood are dimensions-led.

The 'Particulars for a bakehouse, dwelling and lofts on Fore Street, London' produced just a few years later in 1770 is also handwritten on parchment and signed. In this contract the agreement was made between only the master builder and the client, gingerbread maker John Steinmetz. No architect was involved. Rather delightfully, and unusually, three small drawings illustrate the text; an elevation, a plan and a section showing the timber framing.[19] As with the St James Square specification the written description of the building is organized by floors, although these run in reverse from the top of the building to the cellar (this is also the case within '11. Inside' where the highest floor comes first), but in addition some sections for specific trades have crept in:

1.	Dimensions	13.	Chambers
2.	Bakehouse: bricklayers work	14.	Parlour and kitchen
3.	Bakehouse: tylers work	15.	Cellar
4.	Dwelling house: bricklayers work	16.	Stairs
5.	Dwelling house: tylers work	17.	Floors
6.	Bakehouse: carpenters work	18.	Finishing the room
7.	Roof	19.	Plasterwork
8.	Westside	20.	Masons work
9.	Front	21.	Painters work
10.	Carpenters: dwelling	22.	Glaziers work
11.	Inside	23.	Plumbers work
12.	Garret	24.	Smiths work[20]

The Bakehouse specification is dimensions-led but includes elements of the trades-based specification that will come to dominate throughout the 19th and 20th centuries.

The division of the specification by trades is much more explicit in a printed set of documents written one year earlier to specify George Dance the Younger's Newgate Gaol and the Sessions House at the Old Bailey. This specification is divided up into slim folded pamphlets, each one covering the work of the various trades – the Bricklayer, the Carpenter, the Glazier, the Mason, the Painter, the Plaisterer, the Plumber (Fig. 2.2) and the Smith/Ironmonger – with one for each building.[21] Each pamphlet varies greatly in length. The 'Description of the Painter's Work to be Performed in Building A New Gaol in the Old Bailey' comprises of just one clause on the title page while the 'Description of the Carpenter's Work to be Done in Building the Sessions-House in the Old Bailey' runs to eight pages. While the contracts for the two houses would have been let out solely to the master builder, the contract for Newgate Gaol and the Sessions House was let out to individual trades and managed by the architect.[22]

Contracting in gross is usually given as the main reason that traditional specifications changed to the new detailed type of trades-based specification represented by the Newgate Gaol documents. However, the emergence of the general contractor does not become common until the 1830s and there is a different explanation for the use of this new type of specification. The project had been funded with government money as part of what Summerson calls the 'the age of improvement' when efforts were to consider the development of London more holistically and not just leave it 'to the mercy of ignorant and capricious persons'.[23] A number of new Acts of Parliament had made it possible to deploy government funds towards lighting, paving, road widening

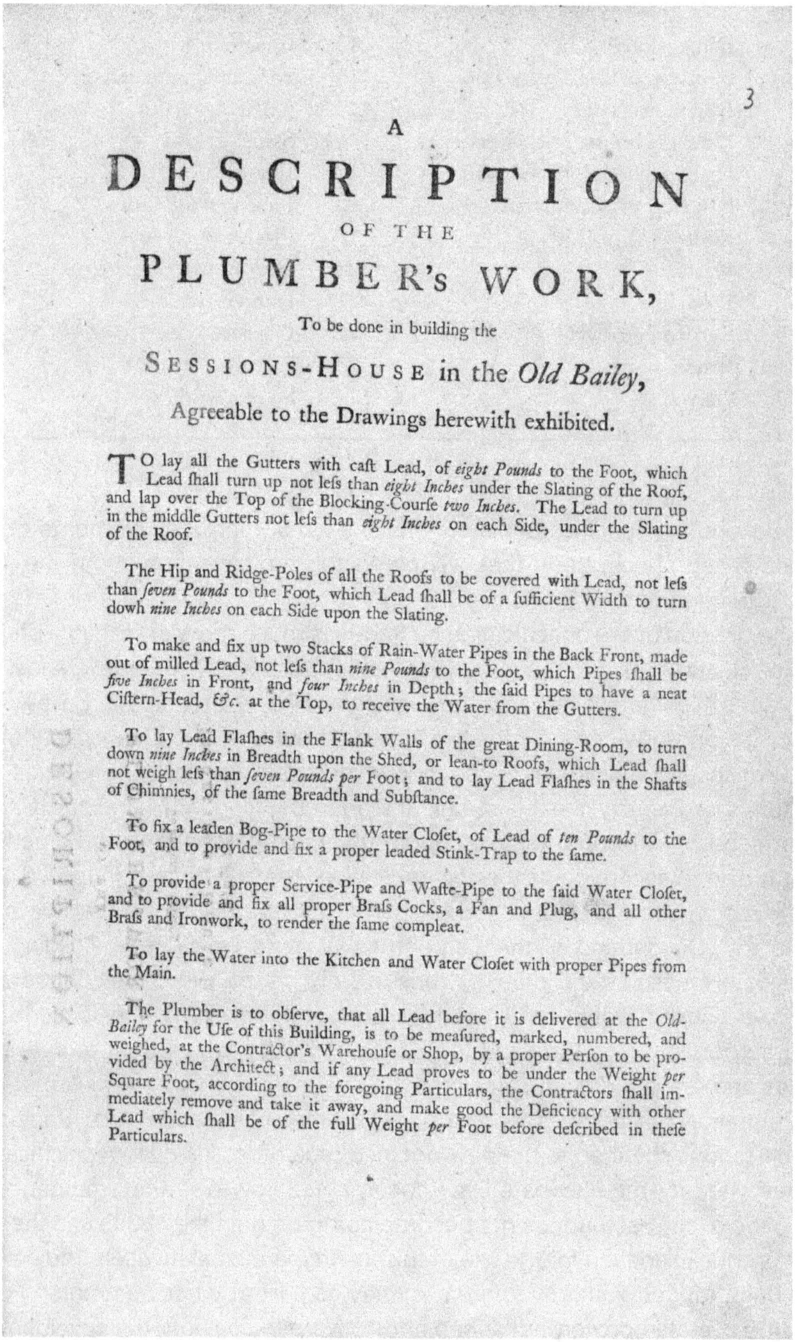

A DESCRIPTION

OF THE

PLUMBER's WORK,

To be done in building the

SESSIONS-HOUSE in the *Old Bailey*,

Agreeable to the Drawings herewith exhibited.

TO lay all the Gutters with caft Lead, of *eight Pounds* to the Foot, which Lead fhall turn up not lefs than *eight Inches* under the Slating of the Roof, and lap over the Top of the Blocking-Courfe *two Inches*. The Lead to turn up in the middle Gutters not lefs than *eight Inches* on each Side, under the Slating of the Roof.

The Hip and Ridge-Poles of all the Roofs to be covered with Lead, not lefs than *seven Pounds* to the Foot, which Lead fhall be of a fufficient Width to turn dowh *nine Inches* on each Side upon the Slating.

To make and fix up two Stacks of Rain-Water Pipes in the Back Front, made out of milled Lead, not lefs than *nine Pounds* to the Foot, which Pipes fhall be *five Inches* in Front, and *four Inches* in Depth; the faid Pipes to have a neat Ciftern-Head, &c. at the Top, to receive the Water from the Gutters.

To lay Lead Flafhes in the Flank Walls of the great Dining-Room, to turn down *nine Inches* in Breadth upon the Shed, or lean-to Roofs, which Lead fhall not weigh lefs than *seven Pounds* per Foot; and to lay Lead Flafhes in the Shafts of Chimnies, of the fame Breadth and Subftance.

To fix a leaden Bog-Pipe to the Water Clofet, of Lead of *ten Pounds* to the Foot, and to provide and fix a proper leaded Stink-Trap to the fame.

To provide a proper Service-Pipe and Wafte-Pipe to the faid Water Clofet, and to provide and fix all proper Brafs Cocks, a Fan and Plug, and all other Brafs and Ironwork, to render the fame compleat.

To lay the Water into the Kitchen and Water Clofet with proper Pipes from the Main.

The Plumber is to obferve, that all Lead before it is delivered at the *Old-Bailey* for the Ufe of this Building, is to be meafured, marked, numbered, and weighed, at the Contractor's Warehoufe or Shop, by a proper Perfon to be provided by the Architect; and if any Lead proves to be under the Weight *per* Square Foot, according to the foregoing Particulars, the Contractors fhall immediately remove and take it away, and make good the Deficiency with other Lead which fhall be of the full Weight *per* Foot before defcribed in thefe Particulars.

FIGURE 2.2 Plumber's Pamphlet, Specification of Works for Newgate Gaol and the Sessions House at the Old Bailey, London (1770). Source: Courtesy of the Royal Institute of British Architects Library Drawings and Archives Collection.

and the erection of bridges and public buildings in London. In this case it is the fact that government money was used to build Newgate Gaol that necessitated stringent accounting procedures and the need to fix the contract sum and the precise details of the work in advance.

A further clue to this early appearance of the trades-based organization of this document is given in a specification produced in 1774 by John Soane's office for Tendring Hall in Suffolk.[24] 'The particulars and estimates of the several works' for Tendring Hall is split into two parts, the contract for the 'shell' of the building that was planned and estimated for in advance (let out to separate trades but still administered by Soane), and the contract for the 'finishing' that was discussed until completion and paid by measurement of the work carried out. Interestingly the particulars of each description have a different structure. The 'finishing' part is organized as a dimensions-led document, in order of the storeys of the building from top to bottom ('Attic Story', 'Bedchamber Story', 'Mezzanine Story', 'Hall Story', 'Great Staircase', 'Basement Story'), whereas the particulars for building the new house and office describe the 'shell' of the house according to trades and in sequence of work (digger, bricklayer, carpenter, mason, plumber, smith, slater).[25]

Soane's specification for the Tendring Hall is a transitional contract, where both the traditional mode of 'contracting by measure' in which the costs of the building were paid by measure after its completion (for the finishes) and the new mode, of preparing the description so that it can be priced in full in advance by the different trades (for the shell) are in use. It suggests that the earlier dimensions-led form of specification – such as the St James Square house – where buildings were described as a dimensioned object, storey by storey, were used when building work was paid by measure. Whereas the trade-based organization of the document as for Newgate Gaol, in which the trades carrying out the work and their tasks determine the arrangement of the specification, emerges when the contractual sum is determined and agreed in advance. In the earlier type, the architect need not be familiar with work on site. Their job is only to determine the form and extent of the building and to supply some details about the materials. Tilo Amhoff suggests in his study of the contractual changes of the early 19th century and their impact on architectural documentation that earlier specifications needed only to give 'literal descriptions' of the building since the craftsmen could be trusted to carry out the work to expected quality.[26] The trades-based specification requires a thorough knowledge of the processes of building and the ability to specify them in detail. Amhoff adds that because these contracts were let out to unknown builders their legal obligations, including the method of work, had to be described in detail. The newer documents replaced a system of trust, on-site negotiation and personal relations. Although the two forms of specification differ because of the contractual arrangements they were needed to secure, they also describe the building in radically different ways, first as finished object and later as a sequence of building processes. This difference is also visible at the detailed level of their forms of clause, as will be discussed with reference to naming in Chapter 3 and to process-based description in Chapter 4.

The Newgate Gaol and Sessions House specifications give much greater detail about the materials and process of building than appeared in the specifications for the St James Square house or the Bakehouse, dwellings and lofts. To some extent this is explained by the buildings' scale and complexity. It is not possible to simply repeat a conventional formula for building as at the St James Square house, and construction details cannot be assumed, but other factors also contributed to the proliferation of building techniques and the need to specify one over the other. For example, Jack Bowyer points out that as the transportation of materials increased with industrialization, buildings were less often made with tried and tested local materials.[27] As a result, a builder might be working with an unfamiliar material and require guidance. Furthermore, the proliferation of architectural styles in the 18th and 19th centuries also added to the increase in techniques a builder might be called on to use.[28] Gelder explains the effects of this multiplication of techniques on the specification:

> If construction is conventional, the specification could be fairly brief – the contractor could reasonably be expected to know enough to fill in the gaps, and the contractors would have a clear enough understanding of normal practice to not require it to be spelt out in the specification. On the other hand, if construction is innovative, the specification will need to be extensive perhaps even being specific about methods and buildability issues that are normally the contractor's domain.[29]

Gelder also explains that literacy rates had climbed enormously by the end of the 19th century with approximately 75 per cent of adult males able to read, making specifications more accessible.[30] But all commentators seem to agree that the most significant change had been the shift from individual craft contracting to gross tendering.[31] Although gross tendering did sometimes take place in the 18th century, and also in earlier periods (according to Gelder the ancient Greeks used this method) it took off in the UK with the London-based large-scale developments of Thomas Cubitt in the early 19th century. In this form of tendering the specification needs to do more than specify one method or material over another, or provide information about an unfamiliar technique. It becomes a tool used to ensure quality of work, that the main contractor or architect deploys to prevent subcontractors cutting corners in order to maximize their profit. According to Bowyer, John Nash foresaw this aspect of the impact gross tendering would have on specification in his contributions to the parliamentary debates of the early 19th century on building contracting:

> John Nash, the architect, was one of the early advocates of contracting in gross. He maintained that the new system would require the architect to prepare a detailed specification setting down all materials and labours. Not only would this ensure that the architect carried out his work properly in the first instance, but also that the supervision of it would be more effective, the clerk of works having a technical document at his disposal stipulating the exact technicalities of the work.[32]

The specification yields an authority that enforces quality of materials and workmanship on site. In the Newgate Gaol specification, it sets out (in the single clause included in Painter's Work section) that the painter is to apply four coats of oil, not three, to all exterior timber and iron surfaces, and to use colours only *as shall be directed*:

Description of the Painter's Work
To paint all the outside Wood and Iron-Works, in every Part of this Building, *four Times in Oil*, and to paint all the inside Wood and Iron Works, in every Part of this Building, *three Times in Oil*, of such plain Colours as shall be directed.[33]

If the specification is comprehensive, it also prevents contractors adding extras to their bills where items were not included. In his 1860 *Handbook of Specifications* Thomas Donaldson explained that in situations where getting the lowest price for a tender became more important than working with tried and trusted contractors, the specification had a particularly important role in mitigating against 'fraud and bad work by unscrupulous men':

> Where Builders of high established character undertake a work, great minuteness of description may not be necessary; but the Architect cannot be too elaborate or cautious, when having to do with a stranger or person of doubtful reputation, as sometimes happens in the case of open competition for public bodies. These boards are less cautious as to the character of the Contractor than anxious to secure the lowest amount; trusting to remedy any defect by the stringency of the contract and elaboration of the specification; forgetful that, even with these precautions, a wide door is still left open for fraud and bad work by unscrupulous men.[34]

This situation of general distrust between contractors and architects has only intensified, but today it results in very different forms of specification such as the open specifications produced by Schumann Smith. Nevertheless, the trades-based specification would continue to be the norm throughout the 19th century and into the second half of the 20th century. The enormous expansion in specification production in the early 19th century is clear to see in the documents from the time. They were often printed and put together with great care; well-organized, detailed, thorough and cross-referenced to the drawings package. Some of the specifications available in the RIBA collection are for private houses and shops but many detail the large public buildings – hospitals, railway stations and asylums – which were new typologies and fuelled the building boom at the time.

It is during the first half of the 19th century that a discourse about specification practice starts to emerge. Soane had already collected copies of his own practice's specifications in their 'Book of Precedents' and the office used these as models for later documents. The prolific architect David Mocatta, who built many of the new railway stations and established a highly commercial practice, collated hand copied duplicates of his practice's specifications and bound them in red leather embossed with the title *Specifications* in gold and donated them to the newly formed RIBA,[35] presumably hoping to showcase his expertise and promote good practice. Later in the century, books giving guidance on specification writing appeared: Bartholomew's *Specifications for Practical Architecture*[36] was first published in 1840 and Donaldson's *Handbook of Specifications* in 1860.[37]

In 1898 a new journal, *Specification*, was launched to give architects guidance on writing specifications (it is still in print today, but its remit is no longer to set out best practice for specification). The first edition of *Specification* was organized according to the order of the trades that was by this time conventional (Fig. 2.3). Although there are 37 trades in this index, the general guidance for ensuring 'a well-written specification' recommends as few as nine categories for most building works. 'In this arrangement', it is explained, 'the trowel-using trades are kept together as are also the hammer-using trades, the metal trades and the finishing trades'.[38] Here the kind of work to be carried out still determines the categories of the specification rather than the type of product or their location in the building. After some general notes on specification the rest of the journal is divided into the trade sections, with some wonderful title blocks showing each craft in progress in rather anachronistic medieval form (Fig. 2.4).

Each of these trade divisions contains general advice about techniques and materials followed by exemplary clauses. The editors are clear that their new publication should replace the inaccessible notes and files of most practitioners and 'become a storehouse of information, readily at hand, invaluable to the

TABLE OF CONTENTS.

Editorial Announcements: PAGE
To Readers—Literary Contributions—
Illustrations—To Correspondents .. 2

Introduction:
Points to ensure a well-written Specification—Note for Subscribers 5—10

DIVISION I. Construction.

General Conditions:
General Notes—Specification Clauses
—The London Building Act 13—17

Excavator:
General Notes—Foundations—Piles
—Specification Clauses—The London Building Act—Glossary 19—23

Well-Sinker:
Shallow Wells—Deep Wells—Glossary 25—29

¡cretor:
General Notes—Materials—Application to various Positions—Specification Clauses—The London
Building Act—Glossary 31—35

Bricklayer:
Bricks—Details and Purposes—Piers—Furnace Chimney Shafts—Workmanship—Specification Clauses—
The London Building Act—Glossary 37—47

Drainlayer:
House Drainage—Town or District Sewerage—Specification Clauses—
The London Building Act—Glossary 49—53

Terra Cotta Worker:
Designing—Materials—Manufacturing Notes—Building—Specification
Clauses—The London Building
Act—Glossary 55—59

Mason:
eneral Notes—Selection—Setting
Masonry—Specification Clauses—
The London Building Act—Glossary 61—69

Pavior:
Plain Paving—Decorated Paving—
Specification Clauses—The London
Building Act—Glossary 71—73

Carpenter:
General Notes—Floors—Girders—
Roofs—Partitions, Linings, and
Fencing—Half Timber Work—
Selection and Detection—Spectator's Stands—Specification Clauses
—The London Building Act—
Glossary 75—87

Joiner:
Materials—Floors—Doors—Windows
—Specification Clauses—The London Building Act—Glossary 89—99

Ironmonger:
Doors—Door Checks—Windows—
Specification Clause—The London
Building Act—Glossary 101—105

Slater:
Slate Rocks—Roofing—Specification
Clauses—The London Building Act,
—Glossary 107—111

Tiler: PAGE
Roofing Tiles—Tables—Miscellaneous
Roofings—Specification Clauses—
The London Building Act—Glossary 113—117

Thatcher:
General Notes—Materials—Workmanship—Specification Clauses—
The London Building Act—Glossary 119—121

Structural Engineer:
General Notes—Structural Castings—
Wrought Iron—Steel—Iron and
Steel Sheets—All-iron Roofing—
Cast Iron Columns and Stanchions
—Iron and Steel Joists or Girders—
Compound Joists or Girders—
Painting Iron and Steel Work—
Specification Clauses—The London
Building Act—Glossary 123—141

Fireproof Construction:
General Notes—Materials—Fireproof
Construction—Specification Clauses
—The London Building Act 143—149

Heating Engineer:
General Notes—Systems—Specification Clauses—The London Building
Act 151—155

Horticultural Engineer:
Conservatories—Glasshouses for
Special Purposes—Miscellaneous
Houses: Heating, Cooling, and
Ventilating—The London Building
Act—Glossary 157—161

Hydraulic Engineer:
Power—Pumps and Engines—Lifts—
Specification Clauses—The London
Building Act—Glossary 163—167

Mechanical Engineer:
General Notes—Buildings—Foundations—Boilers—Specification
Clauses—The London Building
Act—Glossary 169—175

Ventilating Engineer:
General Notes—Natural Ventilation
—Specification Clauses—The London Building Act—Glossary 177—181

Founder:
General Notes—Light Constructional
Cast-iron Work—Grates, Chimney
Pieces, Stoves, etc.—Specification
Clauses—The London Building
Act—Glossary 183—187

Smith:
General Notes—Small Structural
Smithery—Specification Clauses—
The London Building Act—Glossary 189—193

Art Metal Worker:
Ornamental Ironwork—Ornamental
Copper—Specification Clauses 195—197

Zincworker and Coppersmith:
Zincworker—Coppersmith—Specification Clauses—The London Building
Act—Glossary 199—203

Electrician:
Lighting—Specification Clauses—The
London Building Act—Glossary 205—213

Bellhanger:
Mechanical Bells—Pneumatic Bells
—Electric Bells—Specification Clauses
—The London Building Act—
Glossary 215—219

Gasfitter: PAGE
Illumination—Service—Specification
Clauses—The London Building
Act—Glossary 221—225

Plumber:
General Notes—External Plumber—
Specification Clauses—The London
Building Act—Glossary 227—235

Plasterer:
Limes—Cements—Plasters—Internal
Work—External Work—Specification Clauses—The London Building
Act—Glossary 237—243

Glazier:
Glass—Glazing with Putty—Specification Clauses—The London Building
Act—Glossary 245—249

Painter and Decorator:
General Notes—Painting—Varnishing—Gilding—Distempering—
Specification Clauses—The London
Building Act—Glossary 251—257

Paperhanger:
General Notes—Old Work—Specification Clauses—The London Building Act—Glossary 259—263

Blind Maker:
Inside Blinds—Outside Blinds—
Specification Clauses—Glossary 265—267

Furnisher:
Private Building—Public Buildings—
The London Building Act—Glossary 269—273

Gardener:
Entrance Gates, Lodge, and Approach
—Garden Plan—Flower Garden—
Specification Clauses—Glossary 275—279

Road Maker:
Roads and Streets—Footpaths—
Specification Clauses—The London
Building Act—Glossary 281—285

DIVISION II. Professional Practice.

Surveyor:
Dilapidations—Ancient Lights—Mensuration Rules—Tables—Scales of
Fees—Glossary 289—295

Hygienic Engineer:
Water Supply—Sewerage—Specification Clauses—The London Building
Act—Glossary 297—303

Legal:
Some leading Principles of the Law of
Contracts in England and Ireland—
Announcements 305—307

Miscellaneous:
Customs 309

DIVISION III.
Buildings in Progress.

The Surveyors' Institution 312
Colchester Town Hall 313
North Bridge Street, Edinburgh 314
Belfast City Hall 315
Paddockhurst, Sussex 316
The Guildhall, Cambridge 317
Town Hall and Law Courts, Cardiff 318, 319
House for Mr. Julian Sturgis 320

For Advertisers' Directory see Page 8.

FIGURE 2.3 Contents page, *Specification*, 1898. Source: Courtesy of EMAP.

FIGURE 2.4 Joiner title block, *Specification*, 1898. Source: Courtesy of EMAP.

experienced man or the beginner in practice'.[39] Moreover, they explain, the advice it contains should be followed for good reason, because:

> Sometimes the construction of a phrase, or the position or specifics of a single word, may lead to disputes involving the question of many thousands of pounds.[40]

The journal invited readers to send in improved clauses and each edition offered updates, discussion, overviews of 'buildings in progress' and, when necessary, the introduction of new techniques and trades and the removal of redundant ones. As such, it shares features with the current NBS, which Gelder calls a 'skeleton specification' in as much as it sets out a consistent order for the description of the building and a database of exemplary clauses which can be selected to suit each project. The various forms of the NBS are constantly updated in response to changing technologies, standards, regulations and contractual developments, but unlike the journal *Specification* they are commercial products to which practices must subscribe.

The early issues of *Specification* made use of the two techniques which had already enabled specification practice to become more standardized during the 19th century. The first was that examples of good practice were collated and circulated as may have happened with Mocatta's *Specifications*. This was certainly the case with Donaldson's *Handbook of Specifications* which included 46 exemplar project specifications for buildings ranging from prisons to the New Royal Exchange, some as long as 70 pages. The second was the establishment of master specifications. Bartholomew's *Specifications for practical architecture* (1840) was probably the first attempt to do this in the UK. In 1898 Frank Macey put together *Specifications in Detail*, which included more than 500 pages of sample technical specifications for building work.[41] By the 1960s some larger building organizations, such as the London County

Council's architects' office, had put together their own master specification which set out the order and wording of clauses to be used on all jobs still using the trades-based arrangement,[42] but theirs was immensely wordy and included every detail of the work from washing out buckets to cleaning paintbrushes.[43] Despite radical changes in construction techniques and organization most practices were still using the old cut and paste techniques. Just as I learned how to produce specifications for small residential projects in the first practice I worked for in the early 1990s, each new document built on earlier ones, editing them where necessary to suit the project.[44] The wording of a particular clause would be copied over from one specification to the next, and became sedimented. For example, the phrase 'clean sharp river sand' appeared in clauses specifying mortar at the beginning of the 19th century[45] right through until the early 20th century.[46] Whole clauses were repeated from one document to another, even if the reasons for their inclusion were long since forgotten by the specifiers copying them. No one dared omit them in case they contained something important. John Carter describes specifications generated this way as 'progeny':

> In studying job specifications it becomes evident that many of them are the progeny of 'model' specifications written years ago, on which each successive job has left its deposit of 'protectionist' clauses to tie down the next contractor.[47]

Unlike the drawing which starts each time from a blank sheet, these specifications are additive. They get longer and longer, repositories of everything that has been written before.

By the 1960s the trades-based process-oriented specifications that had dominated for 200 years had become so compendious they were no longer effective. For example, The LCC's 'Specification for the Elfrida Rathbone School' (1961) consisted of 161 closely typed pages describing in extraordinary prose-like detail every aspect of the building process – from washing out buckets to the preparation of material samples for approval. The section for the 'Concretor' ran to 15 pages despite the fact that the school was built almost entirely in brick (see Fig. 4.3).

By 1963 the situation seemed untenable and a 'specifications panel' was formed as part of the RIBA technical committee to look at the problems the profession was facing. The process-oriented specification with its 'repetitiveness between the various documents, the danger of discrepancies, and an archaic form of wording and presentation'[48] needed to be replaced if architects were to continue to have a viable role in producing specifications. The work of the panel encouraged intense debate between members, contributors and in the architectural press on a subject usually at the margins of architectural practice and led finally to the launch of the National Building Specification (NBS) in 1973, that would propose a return to describing building in terms of 'ends rather means'. Process-oriented specification, which had seemed for so long the best way to ensure quality of work, and that had at the same time engaged architects with the details of work on site and rendered building in these terms was to become instead a discredited mode of writing architecture to be avoided.

From 'Means' to 'Ends': 20th and 21st century specifications and the variety of forms of clause

In articles and letters published in architectural journals of the early 1960s and in the reports, minutes of meetings and correspondence of the Specifications Panel, a number of reasons were given for the need to change and streamline the production of architectural specification across the profession. In addition to the lack of clarity and length of documents produced by most offices other problems included the lack of integration between the format of the specification and the bill of quantities. Some members of the Specifications Panel already identified potential for the use of computers in the production of building information and argued for a computer-compatible system. But the most serious concern was that specification was moving out of the architect's control and into the domain of quantity surveyors and building systems manufacturers. Allen Ray-Jones saw this as an effect of new 'industrialized building', which produces a 'closed system' in which manufacturers specify their own products and take away any choice from architects.[49] According to Ray-Jones the answer was not to rail against the new methods but to find ways to at least control the way these components are put together, through specification. Others put their concerns more clearly in terms of power relations between the parties:

> The architect has to some considerable extent delegated the specifying and choosing of materials and sub-contractors ... the quantity surveyor and other interested parties including the contractor often avail themselves of the opportunity to influence what materials and whose services should be used.[50]

The recommendation was 'to make the choice of a product to be specified both quicker and easier for the architect'. After an initial interim report, the Technical Information Committee met on 5 March 1964 and agreed to fully endorse the report's conclusion 'that a coded Standard National Specification should be prepared'.[51] Such a national specification was already in use in Sweden but it would take more than a decade of reports and research, and the establishment in 1969 of the new RIBA Limited Company to finance and run the new NBS. John Carter's first report on the RIBA proposal for the NBS was concerned with assembling data and a working group to begin the research, but also with establishing the business case for the new NBS (which he suggested should include a clause that 'obliges the contractor to have a copy' and therefore presumably to pay for one!).[52] Although of course professional concerns for the quality of building guided these developments, the role of the architect was also at stake. As a panel member signing himself only 'Mr George' insisted in his 1963 report to the Specifications Panel:

> The Architect must be the initiator of the specification. He alone must determine the materials, workmanship and components to go into the building.[53]

For the profession, the battle was to retain specification as a central part of the architect's work, and to do so the new NBS was to move away from describing the *processes* of building, towards an emphasis on the building as *object*. As Tony Allot, a member of the team who prepared the first NBS in the early 1970s, explained in 1971:

> NBS policy is based on the belief that specification should state requirements in terms of ends rather than means, but that there are practical limits in most cases.[54]

The effects were visible in the new document. The structure of the first NBS was based on a Swedish classification system 'SfB' that had been adopted in the UK for technical library filing as laid out in the *Construction Indexing Manual*.[55] This system was organized by products and classified each product according to construction 'form' and material. Letters of the alphabet were used as titles for each of the work sections rather than words, and titles referred to products rather than to trades and their work.[56] For example, the 'Concretor' section in previous trade-based specifications was replaced by section 'B' ('Concreting Formwork Reinforcement Concrete finishes'). The section that had been titled 'Painter' or 'Painter's Work' became 'V' ('Painting Film coatings'). The work sections' alphabetical order suggests an immutable logic to the organization of construction in stark contrast to the old system of division by trades that could be added to indefinitely (Fig. 2.5). The order of products and systems still followed roughly the order of work on site, the

FIGURE 2.5 Front Cover, National Building Specification, 1973. Source: Courtesy of NBS.

introduction to the new NBS was adamant that, 'Each work section deals with one main type of product in a defined range of applications. No attempt has been made to identify these kinds of work with 'trades'.[57] The editors explain that this allows contractors to divide up the work sections and distribute it more easily among subcontractors since, 'the division is almost always by kinds of finished work in place e.g. asphalt work, brick work, clay tiling'.[58] SfB was superseded in 1997 by a more computer-compatible system – UNICLASS (United Classification for the Construction Industry) whose letter titles and product classifications still guides the organization of UK specifications today.

Moreover, the long-held techniques of describing methods and techniques of construction were also to be superseded in the new NBS. According to Colin McGregor, another member of the NBS feasibility team who stayed with the company until his retirement, a guiding principle of their work was that 'the specification describes "work in place", that means the finished result rather than the process of achieving it'.[59] This approach was intended to allow contractors to find expedient methods that suited them and to avoid architects' giving instructions that were redundant or even mistaken. This intention certainly led to a reduction in the number and detail of process-based clauses in the NBS, but in practice they were retained (particularly for materials made up on site and to describe innovative techniques) and used alongside clauses that named products and materials to be used, and performance specification, a form of description that was then new to the UK.

What is most notable about the NBS in its various manifestations since 1973 is that despite the rigidity of its overall organization, each work section may make use of more than one form of clause. This is also the first time that commentators recognized and gave significance to the different ways materials and workmanship could be specified, some of which might describe the work in place, with others describing for example, the means of achieving it. It is at this point that a discourse about the variety of 'forms of clause' appears in the literature. Here for example, are Michelmore's tentative classifications of four different 'forms' in his 1963 report to the Interim Specifications Panel:

> Our aim should be to find a form that states once only what is required; where it is required, and how much is required, and also to define how it is to be done and when it is to be done. We must state what we want compactly. It can be in the form of
> a) a performance specification (fence the site)
> b) a particular appearance description (*close boarded* fence)
> c) a description of the specific composition of an item (*softwood* close boarded fence)
> d) a statement of the method of the work (softwood close boarded fence *creosoted before erection*)[60]

This list appears to be based more on Michelmore's observations of clauses in existing documents than on a set of categories in common use. He notes the differences between each of these descriptions without relating them to where they appear or the circumstances of their use. More recently, in a table in *Specifying Architecture* from 2001, Gelder set out his own list of four 'methods' of specifying in which only two – 'performance' and 'construction methods' – directly tally with Michelmore's 'forms':

> Performance (open)
> Descriptive (closed)
> Proprietary (also known as brand or outright)
> Construction methods (also closed)[61]

Gelder divides these into 'open' and 'closed' methods in reference to their contractual force. 'Open' methods are those which leave the final selection of material or construction system to the contractor while in 'closed' methods the selection is determined by the architect or specifier. It is not just that Gelder's list of categories differs from Michelmore's because their content is different. For

Michelmore it is the way in which the fence is described which is decisive, while for Gelder the meaning of the clause is necessarily tied to its contractual context.

This attention to the forms of clause to be used in the specification and the degree to which the use of each kind may be appropriate for particular forms of contract drives the fourth and most recent type of specification included in this study. Schumann Smith started producing bespoke documents in the 1980s for large commercial projects where detailed design would be carried out by specialist subcontractors. They were designed as improvements to the NBS in which performance clauses were the exception rather than the rule. Their specifications were intended to be more up-to-date, more responsive to change and most importantly, more claim-proof than the NBS with its assortment of forms of clause and greater number of closed specifications. According to Tony Brett of Schumann Smith, the variety of forms of clause in use in the NBS was a particular problem they tried to avoid in their own documents.

Instead, Schumann Smith compiled a database that allowed any given material or component to be specified in one of three categories: pure performance specification, traditional prescriptive (or 'closed)' specification (such as naming or the description of methods of work) and a third category – 'descriptive' specification:

A   GENERAL REQUIREMENTS
A.1000   FORMAT, DEFINITIONS AND USE OF THE SPECIFICATION
A.1001   Specification Format

d)   The types of Specification are as follows:
   i)   Prescriptive (P): The section is a detailed materials and workmanship Specification reflecting the Employer's Retained Architect's design solution. The contractor may be required to provide some fabrication details but design responsibility remains with the Employer's Retained Architect.
   ii)   Descriptive (D): The section, when read with the Design Drawings, indicates the visual intent with which the Contractor must comply when undertaking the Detailed Design. The Contractor retains full responsibility for completing the Detailed Design and execution of the works and for meeting the specified performance criteria.[62]

The document overall is set out according to the same alphabetized work sections as the NBS, but the types of description are grouped together by category. Schumann Smith intend the selection of one of these types to be interchangeable. Each form of specification should be applied as easily to one material as to another, and conversely it ought to be possible for any material to be represented in each of these three ways.

As such, the selection of one of these types of clause over another is not supposed to depend on what material is being specified, but on the contractual situation. Where the architect is responsible for an aspect of the detailed design, prescriptive specification is used. Performance specification is appropriate when detailed selections are to be left to a specialist subcontractor. The third and most innovative category, 'descriptive specification', leaves detailed design open to the contractor, but enables the architect to retain some control over the visual appearance of the detail (but not its actual material make up). This type of specification refers to the architect's 'visual intent' and usually refers the contractor back to the drawings.

In practice, however, Schumann Smith's categories of specification were not always interchangeable in this way. Some materials, such as in-situ concrete, resist or demand certain forms of clause whatever the contractual situation. Natural stone, as in the indicative clause for Portland Stone below, still ends up specified through a list of properties, such as geographic source, which would not be included for other materials. Other materials such as plasterboard may be specified without the detail of their properties, perhaps because this information is given in standards outside the specification.[63]

The open specifications produced by Schumann Smith usher in a new phase in the development of specifications in the UK. Where they comprise largely of performance and descriptive specification, as in the Specification for the H. . .y Building (2006), the documents describe the building through criteria for the behaviours of the finished object, and its visual characteristics as given in the drawings. Thus, these specifications are emptied out of the particulars of building processes and names of materials and products that had previously constituted specificity in the architectural specifications of the 19th and 20th centuries. Despite their length and detail (the Specification for the H. . .y Building is 401 pages long) they are more reminiscent of the St James Square indenture than the trades-based specifications that dominated for 200 years, in so far as it is the formal characteristics of the building that once again prevail, albeit overlaid with the constructional and environmental behaviours that secure quality in very different contractual circumstances.

Schumann Smith's tripartite approach to the practice of specification is based on an explicit awareness that different modes of description are appropriate for different forms of contractual organization. This interdependence also explains the major shifts in the format of the specification since the 18th century that have been set out in this chapter (from object to process, and then from means to ends), The overall form of the specification and the kinds and contents of its detailed descriptions are bound to changes in how building is carried out and organized. Factors such as the complexity and scale of building, the degree to which design follows or breaks away from convention and particularly contractual organization effect not only the arrangement of the document, but also the manner and nature of its contents. These factors have always been preliminary to the textual form and arrangement of the documents, but in the UK it was only in the 1960s that the variety of forms of clause was explicitly recognized and simultaneously, the question of which form was most appropriate or expedient emerged within the industry. Hitherto invisible, the form of clause, or how a material is described within the specification, was understood to have implications for the business of building. The position taken here is that, moreover, the forms of clause used in these documents both rely on and constitute radically different concepts of materials, even if, as in the NBS, they are often used together in one document. Just how these concepts arise and their effects on how architects work with materials and understand their capacities will be the focus of the following chapters.

3

Naming Materials

Just as the general format and organization of the specification reconfigures in relation to changes in technology and building practice, variations in how the materials of building are described appear at the more detailed level of the clauses within each document. We turn first to the most obvious way to specify a material – to name it. We would expect to find that naming is the form of clause most common to specifications, and indeed, it is prevalent throughout the periods I have studied. Naming is the sole method for prescribing materials in dimension-led specifications such as in the St James Square indenture (1734). It is probably the most archaic, used for example in the verbal instructions God gives to Noah for the building of his Ark in *Genesis 6.14–16* (which John Gelder suggests that may well have been have been one of the most influential specifications ever written given the Bible's wide distribution and readership!). Like the St James Square document this is a dimensions-based specification, which names the 'gopher wood'[1] Noah was to use:

> Make yourself an ark of gopher wood; make rooms in the ark, and cover it inside and out with pitch. This is how you are to make it: the length of the ark three hundred cubits, its breadth fifty cubits, and its height thirty cubits. Make a roof for the ark, and finish it to a cubit above; and set the door of the ark in its side; make it with lower, second, and third deck.[2]

In trades-based specifications we find materials named alongside other forms of description that give the details of how materials are to be made up on site, the process-based clauses to be explored in Chapter 4. And in the ends-oriented specifications that were formalized in the 1970s as part of the standardization by the NBS, naming is just one of a variety of forms of clause used, including the performance clause. The only exception to the prevalence of naming is in the more recent open specifications, such as those devised by Schumann Smith, where the aim is precisely to avoid nominating or prescribing specific materials in order that their selection is left to the contractor.

It is a complex task to understand how and why one mode of material prescription – whether naming, process-based or performance as the three main types explored in this book – comes to dominate in any one period or document. As we have already begun to see, new forms of contractual organization were responsible for the more radical shifts in the general and detailed characteristics of the specification. Indeed, when Schumann Smith devised distinct categories of specification (prescriptive, performance and descriptive) whose application depended on which party would take responsibility for the detailed design, they assumed the contractual context was the only factor determining how materials are prescribed, and that forms of clause were interchangeable. But in practice this logic breaks down because, to some extent, the adoption of one form of clause over another is also determined by the kind of material being described (which explains the variety of forms of clause used alongside each other in the NBS). For example, process-based clauses have tended to be used for materials made up on site, such as mortars and concrete, and where techniques are new and unfamiliar (in Schumann Smith specifications these techniques are specified indirectly, through reference to British Standards). Performance clauses tend to be used where the material has been performance-engineered, and is already defined by the manufacturer in terms of the behaviours it can achieve, such as strength, insulation or acoustic control. Where a material's behaviours have not yet been tested and quantified or where the properties of one sample of a material cannot be assumed to be identical to another sample, as in the case of quarried stone, performance specification is not possible. In fact, the specification of a naturally occurring material such as Portland Stone, whose appearance and characteristics varies from one quarry to the next, requires naming. Even where naming is to be avoided, as in the open specifications of Schumann Smith, it remains almost impossible to specify such a material without recourse to its petrological name. Indeed, to both name a material and leave its selection open, Schumann Smith devised the peculiar category of the 'indicative material'. The name given refers the contractor to a really existing material, whose appearance their selection must conform to, but withdraws from any commitment to it.

This chapter will show – with reference to the naming of timber – that even in this most widely used form of clause, the kinds of names used vary with changes in the broader context of the building industry and beyond. As will become apparent in the later chapters on process-based and performance clauses, how we come to know and describe materials in architecture emerges out of the practical, technical and contractual realms of building and their social relations. Moreover, as will be discussed in more detail with respect to the emergence of proprietary or 'brand-based' specification in the interwar period in Britain, the mode of naming used can have significant effects. In this case, proprietary specification transformed the role of the architect, and their relationships with their clients and with building products manufacturers. Finally, using the example of a classification system for materials devised to index architects' materials libraries, we will see that even the assumption that one can simply name a material in the specification, and then describe its form through dimensioning or drawing breaks down in some cases and depends on a form/matter conception of building materials that goes back to Aristotle's hylomorphic schema.

Thus we can say, in the light of Foucault's arguments about words and things, that the forms of clause are more than 'words' describing the materials or 'things' prescribed. They are not simply interchangeable, depending on what system of specification is in use; they are part of what constitutes the material's agency. To understand materials in the context of the architectural specification involves the substitution, as Foucault insists, of 'the enigmatic treasure of 'things' anterior to discourse' with 'the regular formation of objects that emerge only in discourse'.[3] Moreover each of these forms of clause also offer radically distinct conceptualizations of materials, and provide a starting point for a theory of materials that challenges the conventional notion of materials as subsets of generic matter. In turn the use of one form of clause over another has effects not only on how the material is conceived, but also on how it can be mobilized in building and on the work and practices of the architect and the other parties involved.

From Species to Brand Names: Changing practices of naming timber

In the early dimensions-led specifications the naming of timber corresponds with the genus and species-based classifications of natural history. Thus, the St James Square indenture (1734–6) specifies the timbers to be used by their common names: wainscoting was to be 'dry yellow deal', 'oak joysts', 'teaked flooring' were to be laid on 'good fir timber Beams' on the Chamber floor, and the doors on the floor where the great parlour was located were to be 'mahogany'. A plan or section could be annotated as simply as this, just with the names of the materials to be used and without the references to quality or methods or work that characterize the trades-based specifications of the 19th and 20th centuries. Although strictly speaking, only naturally occurring building materials can be named by species, manufactured materials such as glass at least follow the principles of classification. The generic name given is given the clause's title such as 'glazing', and the specification through recourse to further subdivisions that resemble the principle of the species. For example, during the 19th century, glass was named with reference to standards ('best seconds', 'good thirds') and also more rarely, by the manufacturer's name ('Ratcliff Crown Glass').

Within the clauses of George Dance the Younger's Specification for Newgate Gaol and the Sessions House at the Old Bailey (1769) compiled just 35 years later and organized in pamphlets by the subcontractor who would carry out the work, materials such as stone are also named (for example, 'Portland Stone' – notably a name that derives from the material's source, rather than its petrological type) and a new practice for naming timber appears, one that continues well into the 20th century:

> A Description of the Carpenters Work To be Done in Building the Sessions-House in Old Bailey
>
> Floors
> To lay all the rest of Floors in every part of this Building, with the best whole, dry, Christiana Deals, free from Sap or Broad Knots, in straight Joints, with planed and tongued heading Joints, and all the Boards are to be Edge-nailed; and to lay clean Deal mitred Borders to all the Slabs of Chimnies throughout this Building.[4]

Quite apart from the new details concerning the quality of the timber and the manner in which it is to be jointed and laid, the deals are named in reference to Christiana – the port where they were traded. By the 19th century this practice had become widespread, as in David Mocatta's Specification for Two houses on Marylebone Street from the 1830s:

> Carpenter and Joiner
>
> The whole of the Fir Timber used in the building to be of yellow crown Memel, Dantzic, or Riga fir, to be sanded free from sap and all large knots . . . the deal to be Christiana or deals equivalent in quality to be well-dried and seasoned and free from large knots shakes and other defects.[5]

The naming of the timbers incorporates the patterns of timber imports from the Baltic, which supported English construction after local supplies of oak became scarce during the 17th century.[6] When sources of supply moved to the 'White Sea Ports' of Murmansk and Archangelsk, and to the west coast of Canada with the building of the Panama Canal and the economic depression in Europe at the end of the 19th century, the naming of timbers reflected these changes, as in Val Harding's specification for his own modernist house at Farnham Common in 1934:

```
Carpenter & Joiner
General. 225. Deals for Joiners work to be first quality
yellow deals perfectly sound and open air seasoned.
Fir. 226. The fir timber is to be from approved White Sea
Ports or from British Columbia.
Deal. 227. The deal is to be first quality Swedish or
Baltic.⁷
```

By this time, there was increased grading and standardization of timber carried out by national timber associations and in the US by the Department of Commerce, and the North American timber industry had begun to develop timber products such as plywood which made use of poorer quality timber and the waste produced by the sawmills. Such materials need not be specified by species. They are themselves more consistent and therefore lend themselves easily to standardization. By 1964, the exemplar specification published in the journal *Specification* names most timbers simply as 'hardwood' or 'softwood' and supplies the information about the sources of timber and their species in a separate table entitled the 'nomenclature of commercial softwoods'.

Softwood Nomenclature

standard name	botanical species	sources of supply	other names commercial or botanical
European larch	Larix decidua Mill	Europe, including British Isles	Larix europæa D.C. Larch (G.B.)
Redwood	Pinus sylvestris L.	Northern Europe and Western Siberia	red deal (red) (G.B.) yellow deal (yellow) (G.B.) Baltic redwood (G.B.) Archangel fir (G.B.) Danzig fir (G.B.) Memel fir (GB.) Norway fir (G.B.)

standard name	botanical species	sources of supply	other names commercial or botanical
			Petchora fir (G.B.) Polish fir (G.B.) Riga fir (G.B.) Stettin fir (G.B.) northern pine (G.B.) Swedish pine (G.B.) Siberian redwood (G.B.) Siberian yellow (G.B.) Siberian yellow deal (G.B.) Kiefer (Ger.) Scots pine (when grown in British Isles)
White-wood	Pica abies Karst	Northern and Central Europe	Picea excelsa Link white (G.B.) white deal (in part) (G.B.) Baltic whitewood (G.B.) northern whitewood (G.B.) Fichte (Ger.) Omorika (Jugo-Slavia) Spruce (when grown in British Isles)

Source: *Specification*, 1964

The *Specification* editors explain that some merchants may still use the names of shipping ports to identify timbers and that the builder may still refer to European softwoods as yellow, red or white deals. Because of this the architect would need to be familiar with the builder's outdated terminology, but should no longer use it in their own professional language. The table makes clear the different modes of naming in use within the building industry at the time, at a moment when a new practice of naming is being ushered in to replace and standardize the variety of practices.

The development of wood products has meant that in contemporary specifications we also see proprietary or manufacturers' names given where timber is specified. The emphasis here is on their status as items to be purchased. For example, in an example from DSDHA's NBS-based Specification for roofing at Sure Start, St Anne's Ward, Colchester (2004)[8] a product name (which should in fact read ThermoWood®) is given for the timber weatherboarding and described curiously in the document as a 'species':[9]

> Section H21
> 115   TIMBER WEATHERBOARDING, EXTERNAL WALLS
>
> –   Boarding:
>   –   Quality of timber (exposed surfaces): To BS 1186-3, Section 4, Class: 2.
>   –   Species: Thermowood (proprietary thermally treated softwood) or equal and approved[10]

Here the natural 'species' of the wood is not what is at stake. Instead, the manufacturer's patent name for their product and the heat treatment that improves its basic properties and characteristics and makes it more suitable for use in external cladding is decisive.

It might seem reasonable that proprietary wood products of timber would be classed as 'species' just as in the case of natural timbers. As in the classifications of natural history, there is the wider class of material 'timber' and then varieties of that 'genus'. Many present-day publications which collate the latest gorgeous examples of innovative uses of materials or offer practical guidance for detailing or constructing with them – *Houses & Materials*, *Construction Materials Manual*, *Material World* 1, 2 and 3, *Timber Construction* and so on – are organized by 'species' or categories of material. They tend to group both natural timbers and wood products under generic headings such as 'wood' or 'timber'. For example, Victoria Ballard-Bell divides her survey of contemporary uses of materials *Materials for Architectural Design*[11] into 'Glass', 'Concrete', 'Wood', 'Metals' and 'Plastic'. The book is concerned primarily with the visible surfaces of building. In the 'Wood' section one double-page spread shows a little box-like pavilion 'Gucklhupf' at Mondsee in Austria (1993), by Hans Peter Worndl, made entirely out of highly manufactured plywood boards that slide open and closed in a multitude of combinations, while another shows a Bamboo Canopy by nArchitects, erected at PS1 Contemporary Arts Centre in New York (2004). Here, freshly harvested bamboo has been kept moist so that it can be gently bent into a woven self-supporting structure. Apart from the fact that plywood is in part constituted of a material that was grown, like bamboo, their properties are very different and there is little in the way that these two materials have been processed or are used in each of these buildings that is shared between them.

A more industry-based manual such as Arthur Lyons' fourth edition of *Materials for Architects & Builders*[12] covers a much fuller range of building materials, acknowledging that their use is more than surface-deep: 'Bricks and brickwork', 'Blocks and blockwork', 'Lime, cement and concrete', 'Timber and timber products', 'Ferrous and Non-ferrous metals', 'Bitumen and flat roofing materials', 'Glass', 'Ceramic materials', 'Stone and cast stone', 'Plastics', 'Glass-fibre reinforced plastics, cement and gypsum', 'Plaster and board materials', 'Insulation materials', 'Sealants, gaskets and adhesives', 'Paints, wood stains, varnishes and colour', 'Energy-saving materials and components' and 'Recycled and ecological materials'. In contrast to Ballard-Bell, Lyons carefully distinguishes stone from cast stone, or timber from timber products such as ply and MDF. He pays attention to the ways materials are produced that in turn affect their properties and potential uses.

While Ballard-Bell's categories suggest that any example of a material could be substituted for another, Lyons' categories take into account a broader set of parameters. Stone that has been reconstituted or timber that has been mulched, mixed with glue and processed to become MDF or chipboard are not, in Lyons' typologies, equivalent to their natural 'ancestors'. A material's defining characteristic may be nothing to do with what it is made out of, but rather, as in the case of insulation, to do with its use in the building. There is a variety in Lyons' index that goes beyond the simple categories of naming, and to some extent, recognizes that building materials are prepared for their role in building with effects on how they are used.

The changes in how timbers are specified until the present day reveal that naming depends on more than genus and species, as it is understood in the classifications of natural history. Even if, as Kant argued, the 'higher concept' of genus is in a logically necessary relationship to species or variety, in the case of building materials what determines distinctions between the 'diverse manifestation' of varieties and more general groupings is not limited to intrinsic physical similarities waiting to be uncovered but also to social, industrial and economic developments. Point of trade or patent comes to be a defining characteristic of a material. What characterizes it may be constituted geographically or legally, as much as physically. The spectrum of names given in the specification are evidence that preparations beyond the facts of botanical species also inform what a building material is considered to be, and bear the traces of some of the (historically contingent) preliminary operations involved in making a 'natural' material such as timber ready for building.

As with other forms of specification, such changes in naming practice do more than alert the keen-eyed researcher to new developments in the building industry. They are not simply evidence of change but can themselves also have an impact on architectural practice, its scope and what it comprises, and in some cases, have a transformative effect on design. A significant example of this is the emergence of proprietary specification in the interwar period in Britain. While the impact of this new practice of specifying materials and products with reference to their manufacture and product name was recognized and discussed by architects at the time, it has not been noted and explored in architectural histories of the period. Now that this practice has become all-pervasive it is more or less taken for granted as a fundamental aspect of architectural specification, as if it was ever thus.

Effects of Changes in Naming: The emergence of proprietary specification

F.R.S Yorke is best known today as a modernist architect, commentator, promoter and secretary of MARS, the British arm of CIAM, but during the 1930s he was also editor of *Specification*, the authoritative guide to best practice in architectural specification which was produced annually in Britain from 1898. Since its inception, *Specification* had shown architects both how to organize a specification in 'divisions' by trade and given in detail exemplary clauses that could be used to specify particular materials within each of those divisions. Each year the publication responded to and reflected changes in the building industry. By 1935 for example, and now using a crisp sans serif modern font, we no longer see a division devoted to the work of the 'well-sinker', 'thatcher', 'smith' or 'bellhanger'. New products and materials were added, as well as new up-to-date formats in which they should be specified.

Yorke's preface to the 1935 edition drew attention to the enormous acceleration in the use of proprietary building products and urged architects to acknowledge this significant development in their practice:

> Proprietary materials are playing a more and more important part in building construction. The early editions of *Specification* ignored them, as architects were inclined to ignore them . . . Today it is impossible to ignore proprietary building products and to continue to practise as an architect.[13]

The massive expansion in the variety of manufactured products available to architects was clearly apparent in this period and it was also a matter of great excitement for designers. Trade fairs such as the Building Exhibition at Olympia in 1936 were attended by thousands of people (with a visit from the King on the second day), and architects took part in designing stands for the manufacturers on show, often in response to competitions they offered. In fact, Le Corbusier's only built design in the UK was a peculiarly shiny, mirrored stand displaying the products of Venesta Plywood Company at the Building Trades exhibition in 1930. These architect-designed stands were illustrated and reviewed as much as the products in the journals of the time and demonstrate a certain co-operation and mutual interest between architects and manufacturers which, according to one anonymous exhibition reviewer in *The Builder*, marked a significant shift in the architect's sensibilities:

We are, moreover, beginning to outgrow the conception of the architect as the superior and infallible Man of Taste, whose whims the manufacturer must always humbly seek to gratify. Industrial design and architectural design are seen to be complimentary, not, as formerly, mutually antagonistic.[14]

Importantly, this edition of *Specification* and the changes to it described by Yorke as, 'the most sweeping and comprehensive in the history of the publication' was that this increase in the availability of products – from new wood products, fibre boards, plasters, concrete surface treatments and paints to cables, switch covers and electric clocks – would also necessitate a change in specification practice. For the first time in the publication, proprietary products would be referred to in the main body of the specification and included within the relevant Trade Divisions. Previously they had all been gathered together in one section at the end of the publication, 'lumped together, irrespective of their nature, in a special section labelled "Building Specialities"' but Yorke's revised format integrated them into the central part of the specification. This apparently minor change marked a distinct transition. No longer were architects just specifying a generic material that could be purchased anywhere and only occasionally referring contractors to named products. They were, as part of their regular practice, prescribing that the product of a particular manufacturer be used. There was much at stake for manufacturers in persuading architects to specify their products. Via proprietary specification the architect had become a conduit for the guaranteed sale of their goods.

Furthermore, Yorke's preface to the next edition of *Specification* in 1936 articulated a problematic similarity between advocating specific proprietary products and advertising. Within the clauses included in the edited trades divisions details of the products were to be kept concise. The proper place for manufacturers to give more detail about their new products was in the 'trade announcement pages' with the result that, 'advertisements may become, in measure, an extension of the editorial pages'. If the job of *Specification* was primarily to recommend good practice in building and detailing through its exemplary descriptions, proprietary specification gave the editors a new role of recommending one branded product over another. This pre-selection of products has since been embedded in the NBS, the most common kind of master specification system in use in the UK today. In fact, when the NBS was digitized, manufacturers could pay to have their products as the default option on the drop-down menu of selections. The 1935 and 1936 editions of *Specification* show us the moment at which the entry of proprietary products into the specification was formally recognized and recommended as best practice.

However, the inclusion of brand names in the specification is not in itself new. As we have seen as early as 1769 the carpenter's specification for the Sessions House at the Old Bailey specifies windows that are to be 'glazed with Ratcliff Crown Glass'. In 19th century documents, product names are more in evidence. Mocatta's specification for the London Fever Hospital (1833–43) refers to Parker's Roman Cement, a product commonly in use at the time.

> The columns pilasters cornices piers division walls of wards &c tinted red or darker in the accompanying drawings are to be executed in cement which is to be Parker's Roman Cement mixed with sand in good proportions.[15]

Thomas Little's specification from 1843 for a house on Princes Street also includes a specification for 'Roman Cement (Parker & Wyatts No 2)'. The manufacturer made more than one kind of cement. The document names other products; 'Bangor Duchess Slates' '4" Redmond's rising butts' and 'Queensware Hopper basins'.[16] But proprietary specification remains relatively rare until the 1930s. It is not a matter of their gradually increasing appearance. Rather, as is clear from Yorke's preface, the inclusion of the notes for proprietary products (and by extension, then, the inclusion of proprietary clauses in architectural specifications) within the conventional trade divisions was a decisive and even contentious acknowledgement that these products were now part of the architect's remit and practice.

That this was the case can be seen in the specification that the architect Elisabeth Benjamin produced in 1936 for 'East Wall' in Gerrards Cross, just outside London. The house is a beautiful and early example of the use of reinforced concrete in domestic architecture in the UK. She and her client Arnold Colaço-Osorio called it the 'St George and Dragon' house. Through the house's centre runs the 'dragon' – an exposed serpentine brick wall whose head encloses the dining area at ground level with a balcony to the master bedroom above, and whose tail encircles the stair. St George, as Benjamin put it, was the 'rigid white concrete structure bestriding it'.[17] Benjamin's specification is peppered with proprietary clauses.[18] For example, the bricks of the central 'dragon' wall were to be 'Southwater Engineering (sewer finals) from Sussex Brick Co. (E29E)'. 'The whole of the plywood is to be obtained from Messrs. Venesta Plywood. Co.' And the fire grate is to come 'from Bell Fireplace Co and build in same (F4)'. The electrical specification is particularly dense with proprietary clauses:

> COOKER.   Cables to be Callenders 7/044.V.I.600meg.Non-Association.
> Cooker Unit to be of M.E.M. [in pencil: G.E.C?] manufacture, vitreous enamelled finish with pilot lamp holder and 3-pin plug for electric kettle. Cooker fuses to be M.E.M.30amp. Double pole in Iron Case. [in pencil: and fix cooker supplied][19]

Contemporaneous with Yorke's preface, there is a notable increase in proprietary clauses in this example, but other project documentation for the East Wall job further demonstrates the implications of this new form of practice. The job files are dense with trade literature, with letters between the architect and manufacturers about their products and the samples they have sent her, and with detailed correspondence with her client about the latest products he has seen and the selections she might make – 'Callenders Cable', 'bleached cork from John Elbow', 'Crabtree' switch covers and so on.

Moreover, like a number of clients who commissioned modernist houses in this period, Colaço-Osorio was himself involved in the building industry. He directed the London branch of the family business, Ripolin Paints, made so famous by Le Corbusier's 'Law of Ripolin' in *Les Arts Decoratifs Aujourd Hui*. According to Colaço-Osorio's sons, he had dreamed of becoming an architect, and commissioning East Wall was a financial extravagance that enabled him to take an active role in design. His attention, not only to the selection of products, but also to the detail of the specification itself was meticulous, and much of his correspondence with Benjamin is concerned with the phrasing and terms of particular clauses as well as with what is specified. Colaço-Osorio supplied Benjamin with the paint specification in full, making use of his company's own products, and it was repeated almost verbatim in the final document:

> **PAINTER.**
> M 1. GENERAL.
>
> All materials to be bought direct the manufacturers Ripolin Ltd. at 3 Drury Lane, WC2
>
> M 1.A DISTEMPER.
>
> Distempter where used to be Duresco and to be applied strictly in accordance with the manufacturers instructions and obtained direct from Messers. John Lines.

M 1.B USE.

All undercoatings and finishings to be obtained ready for use both in respect of body and color. No tinting, thinning, or intermixing except on the specific instructions of the architect.

M 1.C APPLY.

At least one day is to be allowed between application of each coat.[20]

In an interview filmed just before her death in 1999 Benjamin recalls with great enthusiasm how Colaço-Osorio gave her (and Godfrey Samuel who worked with her on the project) 'absolute carte blanche'[21] because he believed that so many designs were spoiled by an interfering client. While no doubt this was true at the level of the formal design, it was through proprietary specification that Colaço-Osorio made his mark on the project. The letter below from him to Benjamin below shows the degree to which he followed the details of the specification and his ongoing habit of rifling through manufacturers' catalogues, checking out the latest products on offer and offering advice to his architect:

3.10.36
Dear Betty,

I understand from our phone conversation that it is not your intention to remove the false ceiling in the hall and entirely lag the domestic hot water pipes at East Wall.
I feel I must write you that clause H27 was inserted in the specification after taking the advice of several persons who had special knowledge of ELECTRIC water heating and in view of our rather excessive sum for this type of heating.
In the circumstances I am sure you will agree that I should be entirely justified in withholding the 10% retention money, if I find the system fails through the specification not having been out to the letter.
[Various other ...]

Yours Sincerely,
A Colaço-Osorio
P.S. Have you been in touch with Crabtree. My catalogue shows the switch covers we want.[22]

Benjamin herself requested copious leaflets and samples from a great range of manufacturers while putting together the detailed design and it can sometimes seem that the range of goods on offer was overwhelming for her. Selecting these goods and keeping tabs on them had become part of the architect's work, as Christopher Powell recognizes in his discussion of these developments in the interwar period:

> One result of the increase in branded goods ... was to intensify the demands made on builders and designers. An understanding of building crafts alone was ceasing to be enough; skills in administration and management were in the ascendant.[23]

If to some extent, Benjamin's client helped her navigate this territory it also appears to have been a site of contestation between them, with Colaço-Osorio demonstrating his technical knowledge and expertise to his young, female and relatively inexperienced architect. Proprietary specification gave a new role to the architect. She was 'shopping' on behalf of her client, and decisions about the selection of products from Smiths Electric Clocks to Magic Robot boilers involved the client in addition to decisions about layout and design, as is still the case today in small-scale architectural work. An ever-increasing array of products that might once have been purchased by the client after the completion of their building is interpolated into the remit of the architectural project, both altering the extent of architectural work and the nature of the client's involvement.

Proprietary specification arose out of conditions that were transforming relationships between architects and industry. Indeed, the anonymous reviewer from *The Builder* went so far as to suggest that the 'modern-minded architect' had become no less than a 'synthesist' between art, science and industry:

> Thus the modern-minded architect will visit the exhibition not in the detached guise of patron (though as such economic function inevitably in some degree, places him) but as an essential collaborator in the fellowship of art, science, and industry in whose hands the technique of building rests. He is the synthesist on whom falls the responsibility for co-ordinating industrial activities.[24]

Proprietary specification enforces these relationships. It gives the architect the role of endorsing and ensuring the purchase of products, and entails for the manufacturer a necessary interest in promoting and developing their goods with architects. Furthermore, the architect and client enter into a new kind of relationship that may remain marginal, or become quite central as it appears to have between Benjamin and Colaço-Osorio. The substitution of traditional forms of naming for branded names in the specification had a substantial role in altering relations between architects and industry, and between architects and clients, and changed the constitution of architectural practice.

At the simplest level, the sudden escalation in the interwar period of proprietary specification as one variety of naming within the architectural specification reveals a marked change in architecture's broader industrial context. It is evidence of the conditions that architects worked within and adjusted to at this time. It draws attention to the plethora of new building products they now incorporated in their projects and the changing nature of their relationships with manufacturing industries. Typically, architectural histories of the period treat architects' interests in these products as preoccupations of their personal design approach, whereas in fact they were necessary negotiations that had to be assimilated within their practice (and the lack of attention to this background in Le Corbusier's case seems particularly surprising given his explicit interest in branded products).

But it is short-sighted to conclude that alterations in the kind of naming in the specifications of the time were little more than symptoms of the increasing variety of materials and products that were arriving on the scene. Yorke's decision to move proprietary specification from its position in the journal *Specification* as appendix or addendum to the main Trade Divisions of the document was more decisive than a mere matter of description. It both signified and produced the incorporation of the selection and prescription of named products into the centre of the architect's practice. It is not just a matter of what name is given, but also of where in the document and in what order these names appear. Conversely, in other examples discussed in following chapters, some forms of clause move out of the main body of the specification into addenda, as the result of other reconfigurations of building practice, with their own effects.

Moreover, these shifts also extend the category of building materials to embrace what would once have been outside the architect's remit – whether technologies such as boilers and electric clocks, or elements such as lighting and furniture. In some cases these elements will also become part of what the architect designs. It may be overstating the case to suggest, with our anonymous journalist, that the architect became a 'new synthesist' between her art and science and industry, but she certainly became at least a conduit, newly imbricated in the promotion of these new products in ways that altered relations both with manufacturers and with her client. If with respect to timber we have seen that new practices of naming are evidence of changes in the supply, selection and ways of working materials, the example of proprietary specification demonstrates that such developments also have impacts beyond the paper limits of the document itself.[25]

Naming and Table 2/3: Materials as varieties of matter

Although naming seems at first a rather straightforward form of specification, a wide variety of factors can inform how materials are named and classified including biological species; branding; where they were sourced (point of trade, quarry etc.); who is doing the naming and even the general properties that determine how they are used (softwood/hardwood). Moreover, how a material is named in the document can have much wider effects on the building industry and on the architect's role, as we saw with the formalization of proprietary specification in 1930s Britain. The naming of building materials is industry specific, and embedded in its practices at any given time, but at the same time naming grants a misleading equivalence to building materials, as if each is a species of a wider genera – here timber, glass or stone – that themselves are simply divisions of material in general.

In setting out his own argument about the relationship between a genus and its varieties in *The Critique of Pure Reason* Kant writes that 'the various species must be regarded merely as different determinations of a few genera, and these, in turn, of still higher genera, and so on'.[26] For Kant, the difference between the numerous salts that chemists of his own time were engaged with studying 'is merely a variety, or diverse manifestation, of one and the same fundamental material'. Variety, in this view, is only the other side of the concept of a genus 'or of any other universal concept'.[27] The simple mode of specifying materials by name lends itself to the idea that any one material can be substituted for another and seems to confirm the notion that any material is simply one of the diverse manifestations of material in general – or 'matter'.

At a conceptual level naming relates to the conventional notion of matter, in so far as it suggests that each material is a particular instance of the more primary category. This, for example, is how Aristotle conceives of materials in Book Zeta of the *Metaphysics* where he gives his account of the form/matter or hylomorphic schema. 'A shape such as a circle', he writes, 'may be imposed on bronze, on stone and on wood'. The pure form of the circle remains the same whether it is made out of one material or another, and conversely, the suggestion is, that each of the materials could receive any kind of form. The material is necessary for any shape-form (or composite), but it is simply substitutable, one for the other. Aristotle's argument is most easily made when he discusses simple forms such as spheres or the house, asking rhetorically for example 'is there a house *over and above its bricks*?'[28] so that we see the house must involve both form (the house-form) and matter (its bricks). But it gets more difficult to make when he moves away from simple forms towards those such as the 'horse-form' where there is a necessary relation between a composite's materiality and its being.

Aristotle's example of the circle, which might equally give its shape to bronze, stone or wood suggests the simplest mechanism of specification (it is discussed in more detail in Chapter 4). There is a form, given either as dimensions in words, as in the case of Noah's Ark and in the indenture for St James Square (1734–6), or more typically in the drawings produced in conjunction with the specification, and there are named materials in which that form is to be realized. Where the specification simply names materials – an ark in gopher wood or beams of 'good fir timber', a roof in Finnish Birch Plywood or a wall cladding in ThermoWood® – they are described as if they are substitutable. Master specifications such as the NBS leave blanks in their clauses where the names of materials can be inserted and in the digitized version of the NBS, these can literally be selected from a list that appears in a drop-down menu.

That this conception of building materials as substitutable instance of matter depends on the notion of form as in Aristotle's form/matter schema is demonstrated explicitly in an example of material classification that was recommended to architectural practices in the 1960s. Table 2/3 (Fig. 3.1) was designed using the Swedish SfB classification system to enable architectural practices to organize technical information in their building product libraries. SfB would go on to provide the structure and categories of for the first version of the NBS in 1973.[29] Like Aristotle's form/matter schema, Table 2/3 makes use of two terms to classify building materials The first category 'Construction Form' (2) runs along the horizontal axis of the table and includes 'forms' such as 'bricks, blocks', 'tubes, pipes' and 'rigid tiles'. The second category 'Materials' (3) runs down the vertical axis and includes material genera such as stone, clay, plastics, wood. Like Aristotle's composites (and unlike the other classifications in SfB) the components of tables 2 and 3 can never be referred to independently. They must always be given in conjunction:

> Table 2 **Construction Form** is never used without Table 3 **Materials**, and for this reason both tables are combined in this section as table 2/3.[30]

In Table 2/3 this results in two curiously abstract general categories, which appear at the end of each axis. One is named 'Any and all materials' or lower case 'y' or and the other is 'Products in general' or upper case 'Y'. They are used to index literature concerned with a material such as clay, but not with any one form of it, or with forms of construction such as bricks, which encompass more than one material (e.g. clay, glass, concrete, etc).

Furthermore, the categories along each of the axes are placed next to each other as if they have the kind of equivalence Aristotle assumes. Along the top the 'forms' loosely reflect the traditional trade divisions of the specification – foundations, structure, walls and ceilings, finishes – a logic from practice that becomes inaccessible once it is abstracted into the elements here. But the

Material		Construction form						
		Cast in situ	Bricks, blocks	Structural units	Sections, bars	Tubes, pipes	Wires, mesh	Quilts
In formed products		E	F	G	H	I	J	K
e	Natural stone		Fe	Ge				
f	Precast concrete		Ff	Gf	Hf	If		
g	Clay		Fg	Gg		Ig		
h	Metal			Gh	Hh	Ih	Jh	
i	Wood		Fi	Gi	Hi			
j	Natural fibre		Fj	Gj	Hj	Ij	Jj	Kj
m	Mineral fibre					Im	Jm	Km
n	Plastics		Fn	Gn	Hn	In	Jn	Kn
o	Glass		Fo		Ho	Io		
In formless products								
p	Loose fill							
q	Cement, concrete	Eq						
r	Gypsum							
s	Bituminous materials							
Agents, chemicals								
t	Fixing, jointing agents							
u	Protective materials							
v	Painting materials							
w	Other chemicals							
x	Plants							
y	Any and all materials	Ey	Fy	Gy	Hy	Iy	Jy	Ky

FIGURE 3.1 Table 2/3, *Construction Indexing Manual*, RIBA Publications (reprinted 1969). Source: Courtesy of NBS.

categories down the side – 'formed' 'formless' and 'agents' – might inspire the kind of 'wonderment' Foucault has described in his encounter with Borges' Chinese encyclopaedia of animals.[31] The building rationale behind the divisions is hard to ascertain even if their poetry is delightful. Table 2/3 is riddled with gaps and anomalies that demonstrate the inadequacy of Aristotle's composite model for use in relation to building materials. Although the structure of the table suggests that, like the periodic table, all positions in the grid could be occupied, in practice there are many columns comprising null

NAMING MATERIALS

(except finishing papers)	Foldable sheets	Overlap sheets, tiles	Thick coatings	Rigid sheets	Rigid tiles	Flexible sheets tiles	Finishing papers, fabrics	Thin coatings	Components	Products in general
	M	N	P	R	S	T	U	V	X	Y
		Ne		Re	Se				Xe	Ye
		Nf		Rf	Sf				Xf	Yf
		Ng		Rg	Sg				Xg	Yg
	Mh	Nh		Rh	Sh				Xh	Yh
		Ni		Ri	Si		Ui		Xi	Yi
				Rj	Sj	Tj	Uj		Xj	Yj
				Rm						Ym
	Mn	Nn		Rn	Sn	Tn	Un		Xn	Yn
		No		Ro	So				Xo	Yo
										Yp
			Pq							Yq
			Pr					Vr		Yr
			Ps					Vs		Ys
									Xt	Yt
								Vu		Yu
								Vv		Yv
										Yw
										Yx
.y	My	Ny	Py	Ry	Sy	Ty	Uy	Vy	Xy	Yy

values. For example, few building materials can in fact take the 'forms' of 'Quilts' or 'Foldable sheets'. Admittedly there are a couple of cases where new technologies such as structural glass have been developed since the table's production and would now fill some of its gaps (such as 'Go'). But these gaps demonstrate the extent to which specific 'construction forms' can only be made in materials with particular properties. Table 2/3 shows us that Aristotle's example of a circle that can just as easily be realized in bronze, stone or wood is not in fact so readily applicable to all forms.

Conversely, Table 2/3 shows that the technique of naming building materials in the table as if there is equivalence between them breaks down in practice, because only some of the construction forms can be achieved with each material. For example, none of the 'formed products' are associated with the categories of form 'Thin' or 'thick coatings', or with 'Cast in situ'. It is only cements, concrete, plaster and bituminous materials such as asphalt that can take these forms. These are the same 'formless materials' that are so often described in the specification with recipes and process-based clauses and here test the logic that is applied to others. A single category of materials 'cement, concrete' can be given as 'Cast in situ', a category which in fact gives no indication of the 'construction form' of the finished object, only of the method of its making. The performance of materials cannot be given at all in Table 2/3 and remains outside its parameters. This attempt to classify the materials of building using the form/matter schema makes it clear that the notion of materials as substitutable is problematic when it is confronted with materials in practice and suggests why a variety of forms of description emerge in the specification, even without the changing contractual contexts within which they are mobilized. The limitations of naming as a means to specify materials in the specification, and the invention of multiple forms of clause explored in detail in later chapters, provide a starting point to argue that more conceptualizations of building materials are at work in architecture's discourses, than that of Aristotle's matter. As we will see, more notions are required for the business of building than that of naming particular materials, as if they are no more than the diverse manifestations of a 'universal concept' of matter.

4

Process

We have seen that the coupled structure of Aristotle's form/matter schema breaks down when it is used as a framework ('construction forms (2) / materials (3)') to classify building materials in Table 2/3. In practice, materials are not as neatly substitutable one for the next as the example he provides of the sculptor's sphere suggests. In fact it is the most matter-like materials in Table 2/3's category 'formless products' – such as cements, concrete, gypsum – for which special one-off categories have to be invented along the 'construction forms' axis, such as 'thin coatings' or 'cast in situ'. Strictly speaking 'cast in situ' is more a process through which concrete is formed, than a form in its own right.

That some 'formless' materials such as mortars, and later, concrete when it starts to be used on the building site in the first half of the 19th century, might better be specified through the process of their making is reflected in a new form of clause that appears in English documents in the later decades of the 18th century. A typical example is found in Thomas Hardwick's Description of the several works to be done in erecting a new chapel and wing buildings in St James Burial Ground, Tottenham Court Road (1791–2). The description given of the methods the bricklayer is to use in making up his mortar is detailed and elaborate:

Brickwork Foundations Top Of Plinth	The joints of all the brickwork of the vaults to be struck and jointed the mortar used in the said work to be compounded of good and well burnt lime made from chalk or stone and mixed with clean sharp grit sand attested to be taken from the river Thames, between Fulham and London Bridge. The mortar to be made up in as small quantities as the nature of the business will admit and to be thoroughly beaten with stampers. All the lime shall be Sieve'd very fine and well chafed slaked under cover and mixed in such proportions as the surveyor shall approve[1]

The document is written in longhand with many crossings out and is not always legible. It will be more typical in later specifications for mortars to find the proportions to be used of lime and sand, but in this early example, they are not given directly (and water is not mentioned at all). The proportions are simply to be approved on site by the surveyor. Nevertheless, some very specific details of the quality and source of the ingredients are given – even the specific stretch of the Thames where the 'sharp grit sand' is to be taken from – as well as the methods of their sieving and mixing. There is also acknowledgement of the bricklayer's judgement in so far as the quantities will depend 'on the nature of the business' and in terms such as 'very fine' and 'well chafed' that are otherwise undefined. These 'process-based' clauses, as I call them, take a very different approach to materials to the practice of naming. Through a clause like this we see that materials do not always arrive on site ready for use, as if constant and unchanging, and available to be shaped to meet the designer's formal intentions. Rather, through a sequence of processes, they are made ready for their use in building.

The listing of ingredients has a long history in the craft manuals that appeared as early as the *Mappae Clavicula*, a collection of 'recipes' for metalwork, dyeing and mosaic that was probably first compiled around AD 600, and exists in a number of manuscripts from AD 800 onwards. In his *History of Techniques* Bertrand Gille refers to this form of instruction as the 'recipe', using the term for a wide range of such descriptions (which in French sometimes took this title). He records examples from treatises covering such domains as agriculture, locksmithing, pottery and metallography. The recipe is, he writes, 'an affirmation, coded or uncoded which permits the achievement of a desired result' and because it can be transmitted in writing and not just by demonstration 'it allows the birth of a technical literature which covers certain domains of material life'.[2] Cyril Stanley Smith and John Hawthorne claim that Theophilius' 12th century treatise *On Divers Arts*, which covered painting, glassmaking and metalwork, was the first of these manuals to give full detailed and realistic accounts of the techniques artisans used in their work. Smith and Hawthorne emphasize the extent to which Theophilius, a highly skilled metalworker himself, moved beyond listing the ingredients and proportions to be used, to detailed descriptions of the processes involved. His intent was not just to educate readers, they suggest, but also to 'share his delight in the use of materials and techniques for making beautiful things'.[3] Gille cites just one architectural example – a recipe for making mortars from Vitruvius' 1st century BC treatise *De Architectura*; '1 part of lime to 3 of quarried sand; 1 part of lime to 2 of sea or river sand'.[4] The instruction does not include any of the processes to be used, only the proportions of the ingredients in the mix. Despite their extensive usage in crafts literature, there are few examples of recipes in English specifications prior to the 19th century. It is not until the trades-based specification comes to dominate in the 19th and 20th centuries, that the process-based clause becomes commonplace.

In order to go beyond tracing the historical emergence of the process-based clause, and ask instead what the conceptual significance of specifying materials through the processes of their making, this chapter introduces a parallel process-based description of the moulding of a wet clay brick drawn from the process philosopher (as Graham Harman describes him[5]) Gilbert Simondon. Simondon's description appears in the first chapter of his book *L'individu et sa genèse physico-biologique*. He takes more than 50 pages to describe both the energetic exchanges through which the brick-shaped mould and clay work act on each other during the process of the wet brick's formation, and the processes that prepare clay and mould for their encounter. In Simondon's hands, this meticulously detailed description of process is used to undo the explanatory power of Aristotle's hylomorphic schema, and to show that the coupling of form and matter is not a sufficient account of the genesis of individuals. In fact, Simondon argues that clay (like Aristotle's bronze) lends itself well to being understood in terms of the form/matter schema because it is already a matter-like material.

The form/matter schema has had a strong hold on the architectural imagination, and many architectural theorists and practitioners have developed critical alternatives to the schema, arguing in particular that matter is not merely the passive recipient form, but has its own agency, even vitality. Simondon is careful in his philosophy to avoid the identification of any kind of 'living' substance per se (whether in his accounts of technical systems such as the brick, pendulum or crystal formation, or living systems such as corals), but he does re-describe the processes which take place during the formation of the brick as clay and mould partaking in a dynamic system. This set of dynamic processes – of form-taking itself – are the ones most commentators pick up and develop, including Simondon's architectural readership. But Simondon also describes a series of processes that make clay and mould ready for the possibility of form-taking, and refers to them as preparations, or more pointedly 'preliminary operations' ('operations préalables'). These preliminary operations are more like those described in the process-based clause. In the clause from the St James Burial Ground specification, for example, it is how the mortar is to be prepared for bricklaying that is set out in detail. The chemical behaviours that will ensue once the bricks have been laid – with the mortar expanding and hardening to fill the space between them, do not need to be described. If the mortar has been properly prepared, this dynamic process will unfold of its own accord.

To further distinguish these dynamic and preparatory processes, and to see the resemblance of Simondon's account with process-based description in the pragmatic context of architectural design and construction, I look at it in tandem with two rather elaborate 20th century specifications for casting concrete in situ. The first of these specifies an early use of reinforced concrete in British domestic architecture, in a modernist house, Egypt End, at Farnham Common (just outside London) designed in 1934 by the young architect Val Harding, a member, with Berthold Lubetkin and others, of the British practice Tecton. The finished concrete at Egypt End is perfectly smooth and painted white. In contrast, the second specifies boardmarked concrete with a pronounced grain, to be used in the assembly hall roofs at the Elfrida Rathbone School for the Educationally Subnormal, built by the London County Council in South London in 1961, with John Bancroft acting as lead architect.

By reading these found process-based descriptions of concrete formed in a timber formwork against Simondon's philosophical demonstration of the processes of clay formed in a brick mould I will first suggest that the process-based clause should be understood as a conceptual alternative to understanding materials as varieties of generic matter in the terms of the hylomorphic schema. Second, Simondon's notion of 'preliminary operations' is important to the theory of building materials presented in this book, because it shifts the focus of attention away from the notion of materials as stuff – as particular instances of matter in general – to a broader conception of materials that includes how they are prepared for their mobilization in building. Even where process is not included in how a material is specified, as in the performance specification explored in Chapter 5, I will argue that the very possibility of defining a material through its behaviours must already have been prepared for during the material's production. The variations in names of timbers that we saw in Chapter 3, often reflected the processes they underwent prior to their arrival to site, ready for building. Although the process-based clause is only one of the forms of clause, and as I will propose, only represents one of the ways materials are conceptualized within architecture's practices, the wider theory of building materials I propose depends on the process-oriented philosophy developed by Simondon.

This 'Process' chapter introduces much of the theoretical ground for the book. As you read the first two sections 'The Process-Based Clause' and '"Nothing But a Transit": Hylomorphism and the forgetting of Process' they are likely to seem somewhat unrelated to each other. The first section looks at the emergence of the process-based clause in English specifications and the second shows in more detail the limits of understanding individuals in terms of form and matter, and how Simondon's philosophy looks instead to the processes of individuation in his critique of hylomorphism. But these philosophical and practical realms come together and inform each other in the two sections of the second half of the chapter, which refer to both Simondon's work and to examples of process-based clauses for concrete casting to better follow and develop his distinct categories of *dynamic* operations in the third section, and *preliminary* operations in the last.

The Process-based Clause

In so far as it describes how materials are to be made up on site, the process-based clause has much in common with the recipe, but it does not appear in English architectural specifications until the latter decades of the 18th century. This is almost certainly due to the shift away from craft-based tendering where expertise in how to achieve 'the desired result' remained within the domain of each trade. On the evidence of the documents themselves, the proliferation of recipe-based specifications is not gradual but emerges quite suddenly and is already prevalent in trades-based from the first quarter of the 19th century when gross tendering is becoming more widespread.[6] Charles Fowler's Specification of the several artificers work for a schoolhouse in Hammersmith (1819) gives a typical clause for mortar, concentrating, like Vitruvius, more on the proportions of lime and sand rather than the processes that Hardwick described at St James Burial Ground (1791–2):

> Bricklayer
> All the brickwork to be of good hard and sound bricks, and no soft bricks to be used in any part . . . the mortar to consist of three fourths of clear sharp sand, and one fourth of good Mersham or Dorking Lime. Well incorporated. No secreted rubbish to be admitted.[7]

Specifications from the 19th century are full of lengthy clauses like these covering many aspects of the mixtures and methods of building and the practice continues in the UK, with increasing intensity of detail until the 1960s. The somewhat standardized specifications produced by the London County Council such as the Specification for the Elfrida Rathbone School (1961) are a good example of these enormous prosed-based documents. Early process-based clauses tend to describe materials such as mortars that are mixed up on site, and in the 20th century they will become the form of clause of choice for in-situ concrete when its use becomes more widespread. But we do see process-based descriptions for other kinds of materials. We have already seen in Chapter 2, for example that the specification for the 'Carpenters Work To be Done in Building the Sessions-House in Old Bailey' (1769) included details of how to lay and fix timbers 'in straight Joints, with planed and tongued heading Joints, and all the Boards to be Edge-nailed' as well as naming the species to be used. It is quite common to see processes prior to those on site referred to, as with the 'clean sharp grit sand attested to be taken from the river Thames, between Fulham and London Bridge' in the Hardwick specification. In just a few cases, process-based descriptions are given for methods of manufacture that occur off-site, as for example in the specification for the London Fever Hospital transcribed by the architect David Mocatta into a bound collection of his practice's documents during the 1830s (and bequeathed to the RIBA) describes how the cast iron columns and girders are to be fabricated prior to their arrival on site:

> Smith & Founders &c.
> All the castings are to be of a perfect character and subject to hydraulic pressure if required. The cast iron to be the best soft grey iron and the wrought iron to be well tempered and hammered. These trades are to leave their works generally in a perfect state *and fit for paint*.[8]

In contemporary specifications other methods such as performance or reference to industry standards are used to ensure the quality of materials produced off-site. Process-based clauses will relate to materials that are being made up on site.

As in Theophilius' long descriptions of building the metal workshop, and of getting tools ready for work, process-based clauses sometimes describe ancillary preparations. For example, Val Harding's specification for his house to be built at Farnham Common (1934) using the then new construction process of in-situ concrete casting includes a clause for the measuring vessels themselves: 'The materials shall be measured in vessels or containers of a nature which make reasonably accurate measurement possible.'[9] In a similar general clause in the specification for the Elfrida Rathbone School (1961) the 'means of measurement' must be approved and mixing must all take place on 'an approved impervious platform':

```
Mixing of Materials
25. In all cases where materials are specified to be mixed
in defined proportions, proper approved means of measurement
shall be provided and all gauging and mixing shall be
performed on an approved impervious platform.[10]
```

Although the process-based clause had been in use in architectural specification for at least 180 years it was only in the 1960s (perhaps due to the excessive length and detail of these descriptions at the time) that it was recognized as a distinct type of instruction. Returning to Mr Michelmore's typology of 'forms' of specification in his 'Report to the Interim Specifications Panel' (1963) buried deep in the archives of the RIBA, he categorizes it as '(d) a statement of the method of work':

> Our aim should be to find a form that states once only what is required; where it is required, and how much is required, and also define how it is to be done and when it is to be done. We must state what we want compactly. It can be in the form of
> **a)** a performance specification   (fence the site)
> **b)** a particular appearance description   (*close boarded* fence)
> **c)** a description of the specific composition of an item   (*softwood* close boarded fence)
> **d)** a statement of the method of the work   (softwood close boarded fence *creosoted before erection*)[11]

Although performance specification was recognized as distinctive form of prescription as soon as it emerged in architectural practice in the early part of the 20th century in the US, the process-based clause seems only to be recognized at the moment when the profession begins to doubt its efficacy. Its identification as a mode of description is simultaneous with the beginnings of efforts – still ongoing and apparently almost impossible to achieve – to eradicate it from the specification, that arose during the long enquiry undertaken by the RIBA in the 1960s and 1970s to develop a standardized specification that would replace the in-house formats that still proliferated in British practice. John Carter's 1968 first report on the RIBA proposal for the NBS still includes 'preparation and assembly techniques' and 'fixing or placing techniques' among his eight categories of information.[12] But according to Colin McGregor who was one of the team who prepared the first NBS in the early 1970s, a guiding principle of their work was that 'the specification describes "work in place", i.e. the finished result rather than the process of achieving it'.[13] Tony Allot, another member of the team, makes the same point in an article from 1971:

> NBS policy is based on the belief that specification should state requirements in terms of ends rather than means, but that there are practical limits in most cases.[14]

It is important, he continues, to avoid 'petty restrictions on the contractor' by describing methods only where necessary and as briefly as possible. This approach was intended to allow contractors to find expedient methods that suited them and to avoid architects' giving instructions which were not useful or even mistaken. Today the NBS consists of a mix of forms of clause, used adjacent to each other, some open (or performance-based) and others closed, including methods of work, although these are much reduced in detail and number than in specifications before the NBS. It is only in the performance-based specifications of in-house specifiers of some architectural practices or consultants such as Schumann Smith that the process-based clause has almost entirely disappeared. To some extent this is only possible because these descriptions are now given elsewhere in industry standards. If the recipe was once transmitted from craftsman to apprentice within a trade, and then took central stage in the specification for two centuries in order to ensure quality control, it is now in the domain of statutory regulation or left up to the contractor to determine as long as the end result meets requirements. In the specification the omission or inclusion of processes of making up materials and construction depends on contractual organization and how quality is ensured. But the omission of process has a conceptual dimension as well. In philosophy it has been recognized as a serious flaw in Aristotle's form/matter schema.

'Nothing but a Transit': Hylomorphism and the forgetting of process

An essay written by Roland Barthes for the catalogue of an exhibition about plastic provides one of the most succinct and memorable accounts of form-taking as a 'magical operation par excellence'. What strikes Barthes first is the matter-like nature of this material. Plastic appears to have no qualities of its own and can take on any form and mimic any other material. Second is the apparently seamless metamorphosis of plastic in the machine described as follows where greenish 'raw, telluric matter' is transformed into a 'dressing-room tidy' as if by magic:

> At the entrance of the stand, the public waits in a long queue in order to witness the accomplishment of the magical operation par excellence: the transmutation of matter. An ideally-shaped machine, tabulated and oblong (a shape well suited to suggest the secret of an itinerary) effortlessly draws, out of a heap of greenish crystals, shiny and fluted dressing-room tidies. At one end raw, telluric matter, at the other, the finished, human object; and between these two extremes, nothing; nothing but a transit.[15]

None of the processes transforming the greenish plastic crystals into shiny accessories are available to the exhibition's visitors. Instead, these operations through which matter takes form are obscured by the machine's casing or 'black-boxed' as Muriel Combes has put it in her commentary on Simondon's work, borrowing a term more commonly used to describe the shrouding of technological operations we cannot understand.[16]

In this short text Barthes renders the forming of the little green 'tidies' in the terms of the form/matter (or 'hylomorphic') model that is as familiar in architectural discourse as it is in philosophy. Like the white 'high performance composite' zp®150 that is fed into 3D printers to emerge in any complex shape a designer might determine for it, this greenish 'raw, telluric matter' appears to undergo an 'effortless' transformation into its final form. The main elements of the hylomorphic schema are usually associated with Aristotle's classic account in Book Zeta of the *Metaphysics* where he identifies 'matter' with a material – bronze – and distinguishes it from the 'shape-form' or geometry which in conjunction with it constitutes the 'composite' or 'whole entity':

> In speaking here of matter I have in mind, say, the bronze of a statue, while by shape-form I mean the geometry of the object's appearance and by the composite the statue itself as a whole entity.[17]

According to the schema any entity is comprised both of form – here 'shape-form' – and of matter – whether Aristotle's bronze as here, Barthes' plastic, or Descartes' wax.[18] Form and matter are distinguishable but not separable and in Aristotle's discussion, what gives them the appearance of separability is their substitutability. A shape such as a circle, he writes, 'may be imposed on bronze, on stone and on wood'. The pure form of the circle remains the same whether it is made out of one material or another, and conversely, the suggestion is (as we saw in Table 2/3) that each of the materials could receive any kind of form – whether circle, square or dressing room tidy.

Aristotle does not in fact ascribe any active individuating power to form, or render matter passive in his account despite the fact that he describes the circle as 'imposed'. According to him, it is a third 'something' that generates 'the production':

> Anything that is produced is produced by something (and by this I mean that from which the origin of the production comes), and it is produced from something (let this not be the privation but the matter – we have already defined in what way we are speaking of this), and it is produced as something (i.e. either a sphere of a circle or whichever it might be of the other figures). And just as the output of the production is not the substrate, bronze, so also is it not the sphere, except accidentally, in so far as the bronze sphere *is* a sphere and it is a bronze sphere that is produced.[19]

It is unclear whether 'produced by something' refers to the intention or plan to produce (later he will refer to the intention which determines the building's coming in to being or 'thought' as producing a building[20]) or more literally to the mould or mechanism which brings it about, but it is certainly not form that is prior or generative since form is clearly distinguished from this, as the 'sphere or circle or whichever' is produced *as* something. But in later versions of the schema, and particularly in those that are subject to contemporary philosophical and architectural critique, much is made of the supposed individuating power of form that hylomorphism assumes. Here, for example, is Gaston Bachelard:

> I was immediately struck by the neglect of the material cause in aesthetic philosophy. In particular it seemed to me that the individualizing power of matter had been underestimated. Why does everyone always associate the notion of the individual with form?[21]

Like so many architectural commentators who are concerned with the emergent form-finding properties of materials, Bachelard responds to hylomorphism's supposed privileging of form with a call to address the 'individualizing power of matter'. This is not the problem that arises for Barthes, nor as we will see, is it a direction that Simondon takes in his extended critique of hylomorphism.

The correlation of form with geometry or shape that is often made much of in architectural discourse is not central for Aristotle. Aristotle uses the ordinary Greek word for timber 'hyle' (the first instance of this usage) to refer to matter. He understands it as that which is correlative with form in any substance but is not changeable or 'accidental' in the same way.[22] Aristotle includes geometrical form among the range of possible forms, but 'horse-form' and 'Socrates-form' are there too. Because these cases are always made of the same stuff – 'in flesh, bone and the familiar parts' – there is not the same potential for substitution as in the case of the sphere. As a result, he points out, we are less inclined to imagine that the substance of the horse or man is divisible into matter and form. It is not until the 17th century, when Descartes defines matter as extension, that the concept of form (still paired with matter) is understood in terms of ideal geometry and related back to the Platonic forms,[23] in the sense that will be so crucial for the matter/form binary as it appears in architectural discourse.[24]

Barthes goes on in his essay to be troubled by the way plastic no longer manifests itself as a specific material or as a resource 'still of the earth'. In part this is an effect of its apparent infinite mutability into an endless array of possible forms, and of the apparently sudden nature of its transformation. What are obscured in Barthes' magical plastic mutating machine are the processes of this metamorphosis – of matter-taking-form. Visitors to the exhibition see only input and output as if there is nothing taking place in-between. For Simondon it is precisely this 'veiling' or 'forgetting of operation' as Combes has put it,[25] that is a central problem of the hylomorphic schema, and moreover of its far-reaching effects. Indeed, in 'Form and Matter' he tells us in words that vividly recall Barthes' description that, 'The hylomorphic schema corresponds to the knowledge of a man who stays outside the workshop and only considers what goes in and what comes out.'[26] What the man outside the workshop, observing only seeing the raw material as it goes in, and the finished output (or 'terminal form' to use Barthes' delightful phrase) when it emerges, fails to see or understand are the chains of operations through which form-taking in fact occurs. For Simondon, as Adrian Mackenzie has summarized:

> The basic problem with the hylomorphic scheme is that it only retains the two extreme starting points – a geometrical ideal and formless raw material – of a convergent series of transformations, and ignores the complicated mediations and interactions which culminate in matter taking-form.[27]

For Simondon this is more a problem of how and what we choose to know, than a problem intrinsic to specific processes of form-taking or a call to restore the 'individualizing power of matter'.

On the one hand, as Simondon makes explicit in 'Allagmatique', a dense text appended to the full text volume *L'individuation à la lumière des notions de forme et d'information*, his emphasis on process or 'operation' (the term he prefers to use) is part of his 'allagmatic' method and theory.[28] His broad philosophical aim is nothing less than to give an alternative 'theory of operations' which is symmetrical to the 'theory of structures' which he suggests, most sciences and philosophies concern themselves with. Allagmatic theory can, claims Simondon, reveal operations such as energetic transformations and changes of state that would otherwise go undiscovered and un-theorized. On the other hand, *L'individu* is primarily concerned with giving an alternative account of individuation. For Simondon, as we will see, it is a mistake to look to the already individuated individual to try to understand its coming into being. First, there is a problem insofar as the hylomorphic schema locates the principle of individuation in terms prior to the individuation:

> A term itself is already an individual, or at least something capable of being individualised, something that can be the cause of an absolutely specific existence (haecceity), something that can lead to a proliferation of many new haecceities. Anything that contributes to establishing relations already belongs to the same mode of existence as the individual, whether it be an

atom, which is an indivisible and eternal particle, or prime matter, or a form.²⁹

This is a logical problem because another individual has to be posited to account for the individuation. Second, this account of individuation fails to take up the possibility that the individuation itself could furnish its own principle. It assumes that the principle of individuation must lie outside the individuation in either the matter or the form. Simondon proposes that we look instead to 'the genesis itself as it is in operation' as 'the veritable principle of individuation':

> The veritable principle of individuation is the genesis itself as it is in operation, that's to say the system in the course of becoming, during which energy is actualized. The true principle of individuation cannot be found in that which exists before the individuation takes place, nor in what remains after the individuation is accomplished; it is the energetic system which is individuating in as much as it realizes in itself this internal resonance of matter taking form, and is a mediation between orders of magnitude. The principle of individuation is the unique manner which the internal resonance establishes itself from *this* matter in the taking of *this* form.³⁰

The principle of individuation is not some kind of blueprint that exists prior to the process and gives it some particular form or other, nor is it to be found in some kind of self-propagating force intrinsic to the 'matter' itself. Rather it is the energetic process of becoming a system. In order to explore the individuation of '*this* matter in the taking of *this* form' Simondon devotes his first chapter in *L'individu* – 'Form and Matter' – to a long and involved re-description of the moulding of a brick. While some forms of genesis, such as the growth of the crystal in a supersaturated solution (the central example in 'Form and Energy', the second chapter of '*L'individu*) lend themselves easily to being understood in terms of a dynamic individuating system, the moulding of the wet clay brick resists this kind of description because it seems so perfectly to exemplify the hylomorphic schema. Simondon seeks to show that the schema is inadequate, even in this apparently paradigmatic case of the 'magical' transmutation of matter. There are always chains of operations involved in form-taking. Rather than attempting to understand individuation deductively, by looking at the brick as an already individuated individual, or even at the clay and the mould, we are to 'try to grasp the entire unfolding of ontogenesis in all its variety, and *to understand the individual from the perspective of the process of individuation*'.³¹ In painstaking detail he relates the transformations, mediations and interactions – which culminate in the forming of a wet brick.

Importantly, Simondon proposes that these operations must be understood at a variety of scales. It is not just a matter of going into the workshop to observe what happens there. Indeed, the worker is often unaware of the fullness of these operations, believing him or herself to be bringing about the 'transmutation of matter'. He writes, instead, that we are to 'go into the mould':

> To know the veritable hylomorphic relation, it is not even enough to go into the workshop and work with the artisan: one would have to go into the mould itself to follow the operation of form-taking at the different scales of physical reality.[32]

Through 'going into the mould' with Simondon we become aware of the molecular processes of the clay, as the energy that has been stored up from all the kneading prior to being packed into the mould, is released and propagates, molecule to molecule, only to be limited by the form of the mould. Simondon will call these processes that unfold during form-taking 'dynamic operations' but furthermore, he asks us to look at the processes that prepare the clay for form-taking from its extraction from the ground to its industrial preparations. To a lesser extent, he also explores the preparation and wetting of the wooden mould. These 'preliminary operations' are distinguished from the operations of form-taking, in part because they are prior to them, but also because they might operate at rather different scales. In the following somewhat enigmatic extract, the micro-scale of the 'molecular inter-elementary interactions of aluminium hydro-silicates' is contrasted with the macro-scale of 'working, workshop, press, clay' that the idea of the 'mould' stands in for:

> The mold, thus, is not only the mold, but the technical term of the inter-elementary chains, which comprise vast sets locking up the future individual (working, workshop, press, clay) and containing potential energy. The mold totalizes and accumulates these inter-elementary relations, as prepared clay totalizes and accumulates the molecular inter-elementary interactions of aluminium hydro-silicates.[33]

Preliminary operations prepare what Simondon calls the 'preindividual' state of any system of individuation. They directly prepare the potential of the system to individuate, which gives rise to the dynamic operations of individuation without being directly instigated by the preparations. Simondon explores a number of systems of individuation; 'technical individuation' as in the case of clay moulding; 'physical individuation' as in the case of a swinging pendulum; 'living individuation' in the case of a single cell organism or 'psychic individuation' as in the case of members of a human collective. The preindividual state is common to all of them, but the preliminary operations that prepare the preindividual are only discussed in relation to the moulding of the clay brick. They appear again in the conclusion to his study of technical objects *On the Mode of Existence of Technical Objects* suggesting they are characteristic of technical systems. Unlike living organisms for whom the preindividual state arises through naturally occurring (un-designed) processes, technical systems are prepared with a 'form-intention' as he puts it, whose coming into being is made possible through preliminary operations.

The detail with which Simondon explores the operations of the individuating systems he studies is one of the most immediately remarkable characteristics of his work and method. These intense encounters with the specific constellation of energetic and structural exchanges across a range of examples, and Simondon's attention to the differences between them, are what enable his thought to emerge. As with all the descriptions that follow in *L'individu* and in his other texts, the long account of the moulding of the wet clay brick is both an account of what happens that might resemble those of the physical sciences, and at the same time, the unfolding of a process-based alternative to a theory of structures. In this sense, we might say that his descriptions are themselves prepared towards the demonstration of his own ontogenetic philosophy. In the examples that follow in *L'individu* the emphasis will be on the micro-scale of energetic and structural exchanges. These dynamic operations of individuation itself are what will concern him primarily, and are what are taken on by his commentators from philosophers such as Deleuze and Massumi, to architectural theorists such as Lars Spuybroek. They are veiled even to the artisan who prepares the clay and moulds it – and it is Simondon's descriptions that enable us to go, with him, 'into the mould' where they become visible and at the same time available for thought.

The preparations of the clay (and mould) for their encounter are, however, much more readily accessible. Even if we usually bracket off these chains of operations and exclude them in our accounts of form-taking, preferring to see only the moment of matter-taking-form, the preparatory processes are known to us. When we bake an apple tart, we know something about the long chain of processes that brought the flour to our table, from the development and nurturing of the right strain of wheat to its grinding to the right consistency for making our pastry. We know that the butter will need to be hard but malleable, and that as we mix it with the flour we must be careful not to let the heat from our fingers and their movements melt it too soon. We must let the dough 'rest' so that it will become sufficiently elastic for rolling and shaping and prick it and part-bake it with dry beans so that the base is flat enough and sides high enough to form a receptacle for the apples. All the time, without considering the molecular operations and probably not understanding them (and certainly not considering ourselves as preparing a succession of preindividual states!), we are controlling when and how we introduce energy into the system, and taking measures to edit out the system's own unwanted and self-propelling behaviours, guiding them towards an intended outcome.

Preparations such as these are often described in detailed verbal and textual prescriptions that have much in common with Simondon's account of preliminary operations. Centuries of experience and trial and error are accumulated in the recipes we make use of, (in the case of pastry, the techniques are thought to have been known by the Ancient Egyptians, but the earliest recipes we have that include more than lists of ingredients date from the 16th century) and moreover the 'form-intention' is embodied in the already-prepared flour and pastry, in the tin case and in the design of the oven. For technical operations, such as glass blowing or silver-smithing, there is a long history of describing these processes in language and often in the kind of exquisite detail we see in Simondon's account. They are sometimes attended to in literature as in Primo Levi's accounts of the rigger's work in *The Wrench* or of his own work as a chemist in *The Periodic Table*, but more commonly in the craft manuals that Gille examines in his discussion of the 'recipe'. Like the process-based clauses in the architectural specification, these are *found* descriptions of some of the processes Simondon puts into words in such exquisite and painstaking detail in the 'Form and Matter' chapter. As reflects the requirements of practical knowledge, these found descriptions are more concerned with the preparations for form-taking than with the dynamic processes that are set in motion by them at the micro-scale.

Notable in Theophilius' descriptions in his *On Divers Arts* is the extent of the preparatory processes he saw fit to include in his guidance to artisans. For example, in his third book, *The Art of the Metalworker*, it is some pages before he arrives at instructions for refining and casting silver and gold. Having made his dedication to God, he begins far back in the chain of processes with a section on how to construct the workshop! The building is already informed by the processes it is to house, with a certain orientation eastwards and windows at a certain height and as many 'as you wish and are able to' put and rooms to separate casting operations, and the various materials to be worked (gold and silver get their own rooms).[34] He describes how a trench is to be dug and lined with timber to provide a seat and table for the workmen, how the forge is to be built and how the wall next to it is to be lined with a mixture of kneaded clay and horse dung 'so that it will not be burnt by the fire'.[35] There are sections on all the tools that must be collected – anvils and stakes, hammers, tongs and pliers, iron plates and the 'organarium' for preparing wire, engraving tools, scrapers, burnishers, punches and so on – and in some cases, as for files, he describes how they are themselves to be prepared – sprinkled when red-hot with the ground powder of burnt ox horn and salt – so that they will be hardened for the job. In Chapter 23 on refining silver, we catch sight of the molten silver's own processes, when Theophilius warns the reader that if we see it 'boiling and jumping out' it is sign that there is still some tin or brass in the mixture. But usually, Theophilius concentrates on what it is the workman must do.[36] When it comes to casting itself, the section is very brief. All the detail has been in 'arranging these things' for the process of form-taking. This long and various chain of operations prior to casting, have prepared the silver and mould, and the source of energy-giving heat for the possibility of form-taking which finally unfolds through its own dynamic processes, un-described and without the workman. In this sense, the process-based descriptions for casting concrete in situ have much in common with Theophilius' artisanal guidance. They too focus on getting concrete (and the formwork) ready for casting, so that the process of individuation – the dynamic encounter between mould and liquid stone – unfolds out of sight, of its own accord.

Dynamic Operations in Process-based Description

Simondon's examination of the micro-operations of form-taking in process demonstrates, as he explains in the following extract from *L'individu*, that the preparations of clay and mould have done no more than set up the potential for a series of energetic exchanges to unfold and that the operations of form-taking are, finally 'self-actualizing' rather than imposed on the clay by the mould or by some other external force:

> The principle of individuation of brick is not the clay, nor the mold: this heap of clay and this mold will leave other bricks than this one, each one having its own haecceity, but it is the operation by which the clay, at a given time, in an energy system which included the finest details of the mold as the smallest components of this wet dirt took form, under such pressure, thus left again, thus diffused, thus self-actualized: a moment ago when the energy was thoroughly transmitted in all directions from each molecule to all the others, of the clay to the walls and the walls to the clay: the principle of individuation is the operation that carries out an energy exchange between the matter and the form, until the unity leads to a state of equilibrium.[37]

As such, one of the notions often considered inherent to the hylomorphic schema – that form acts on a passive matter – is challenged. Strictly speaking this was not set out in Aristotle's version of the model where matter is not so much as passive as that which is correlative with form in any substance but cannot bring about any change.[38] And as we have seen for Simondon, the emphasis is more on a shift towards a process-oriented or allagmatic philosophy than on recognizing matter's active potential. Nevertheless, this has been a central concern and opening for many of the philosophers and architectural theorists and practitioners who have been critical of hylomorphism. Manuel DeLanda, for example, proposes that we might recognize active potentials that 'come from within materials' and that we might find alternative material systems that would allow materials to 'have their say in the structures we create':

> We may now be in a position to think about the origin of form and structure, not as something imposed from the outside on an inert matter, not as a hierarchical command from above as in an assembly line, but as something that may come from within materials, a form that we tease out of those materials as we allow them to have their say in the structures we create.[39]

We can see this intention in the work of a practice such as NOX where material systems such as the hanging wet wool strands that were used in the design phase for their World Trade Centre entry 'Oblique WTC' (New York, 2001), are set up that can find their own forms, and are then (with difficulty) scaled-up and translated into other material vocabularies that rarely have the same properties as the starting experiments. So, for example the small-scale stretched latex experiments that were used to develop the design for 'Soft Office' (Stratford-Upon-Avon, 2002) were to be realized as partitions and ceilings made with ply shells that follow the forms that emerged from the latex but have nothing to do with its material and constructional logic. There are also examples of emergent material systems with no translations, and some of the most compelling are the range of techniques involved in concrete fabric-formworking, that can be seen in the work of Miguel Fissac in Spain (but only at the level of cladding), and more recently in the extraordinary work of Mark West and his colleagues at CAST (making columns and beams in particular), among many others including Alan Chandler and Remo Pedreschi (who work primarily at the scale of the wall). Here conventional rigid formwork is replaced by a flexible formwork that is only able to stiffen and behave as a mould once the liquid concrete has been poured in and its weight puts the fabric mould into tension.

In Chandler and Pedreschi's 'Wall One' (UEL, 2004) an adjustable 'jig' holds the fabric mould in place during the pour and allows for the fabricators to manipulate the form throughout the process of pouring and casting (Figs 4.1, 4.2).[40] Furthermore, the ruptures and changes in colour between pours, the micro-landscape of variegated sizes of aggregate as they push to the surface and the

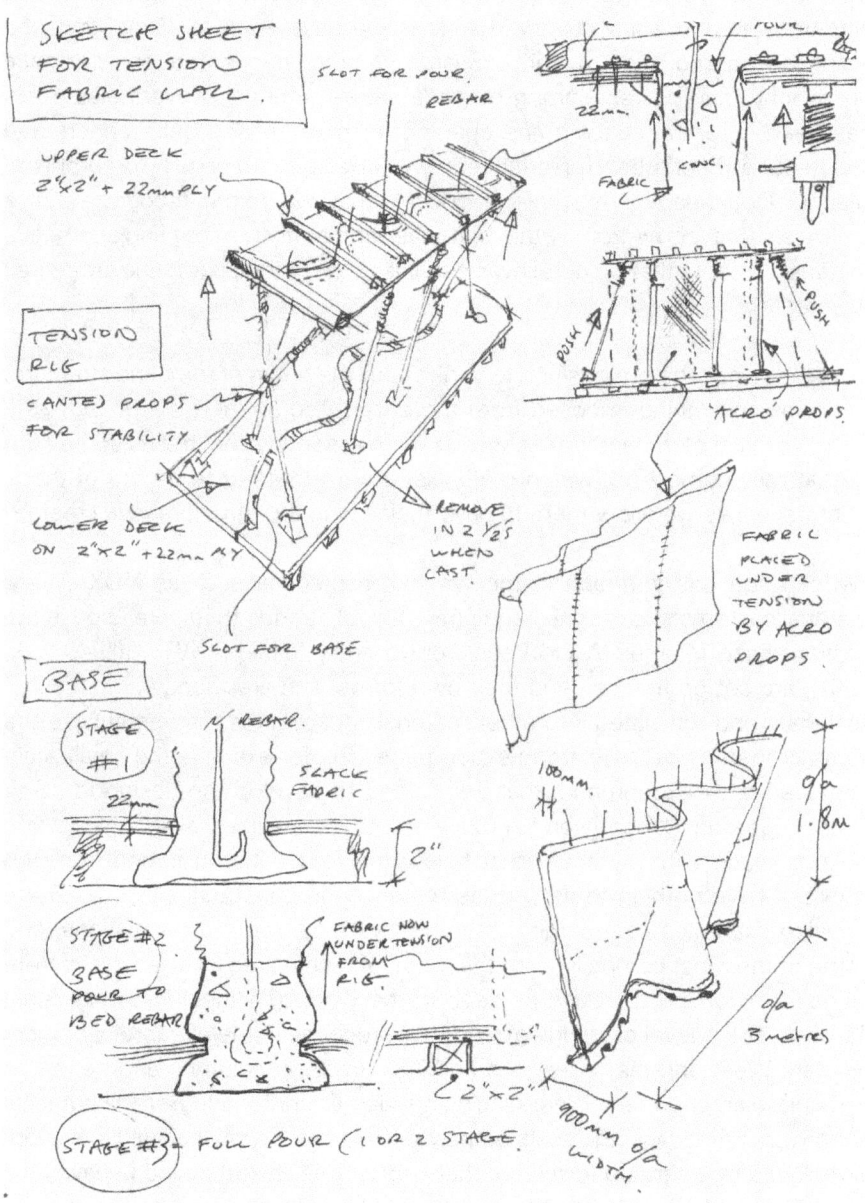

FIGURE 4.1 Preparatory sketch for Wall One showing jig construction, Alan Chandler (2004). Source: Courtesy of Alan Chandler.

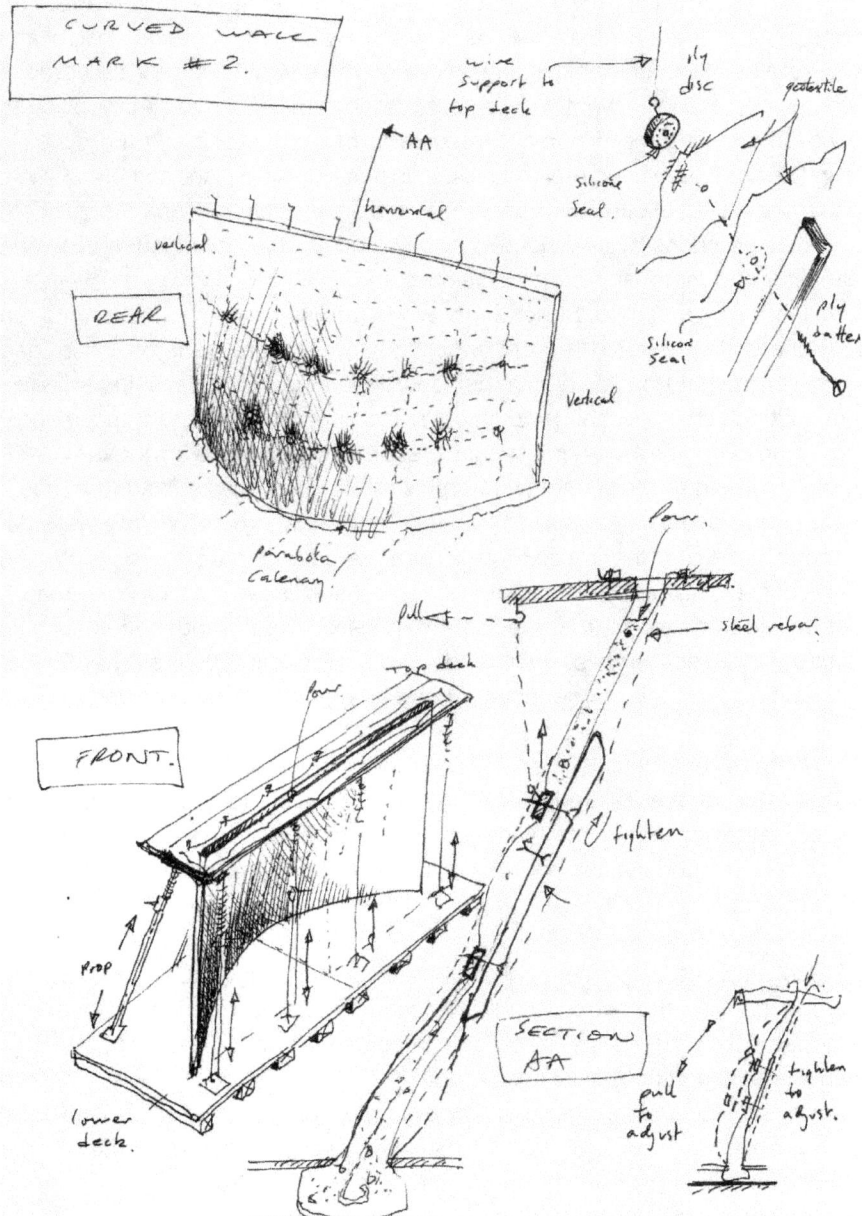

FIGURE 4.2 Preparatory sketch for Wall One showing concrete pour and tensioning process, Alan Chandler (2004). Source: Courtesy of Alan Chandler.

pock-marking of water sweating out as the concrete sets are also left registered in the surface of the finished wall. As such form is emergent at both macro- and micro-scales. If the energetic interactions between mould and clay are veiled in brick forming and in conventional concrete casting but nevertheless take place, in fabric-formworking the drawn-out temporality of the process and the dynamic operations between concrete, mould, the rig that holds it in place and the workers who manipulate, respond and intervene in the process are all highly visible.

For some critics of hylomorphism, this kind of changed relation between fabricator and material is central. Deleuze and Guattari's *A Thousand Plateaus* is full of references to Simondon's work (though only to the writing on individuation and not to his work on technology) that are, as is typical in their work, somewhat transformed in their appropriations of his ideas. In particular, they distinguish 'artisanal' production as an alternative to the imposition of form (and fabric-formworking would presumably be a powerful example of this). According to them, the hylomorphic model 'assumes a fixed form and a matter deemed homogeneous'.[41] Most notably what are 'left by the wayside' in the schema are 'the implicit forms' of a material that are more visible and unavoidable in natural and variegated materials such as wood (and presumably, less visible in materials that are more matter-like and homogeneous such as concrete, plastic, bronze, wax and wood products such as MDF that reconstituted to be homogeneous). These 'things' they write, include:

> An entire energetic materiality in movement, carrying *singularities* or *haecceties* that are already like implicit forms that are topological, rather than geometrical, and that combine with processes of deformation: for example, the variable undulations and torsions of the fibers guiding the operation of the wood.[42]

To take account of these implicit forms is, they suggest, 'A question of surrendering to the wood, then following where it leads by connecting operations to a materiality, instead of imposing a form upon a matter.'[43] This would mean, letting the irregularities of a piece of wood have a role in determining its shape, rather than, for example, subjecting it to 3D carving, an approach that architectural theorist Pauline Lefebvre observes causes some anxiety for a group of architects designing a reception desk in timber, who rather hope to follow, 'what the wood wants to do'.[44] She argues that this goes beyond architects' modernist concern with 'truth to materials', citing Deleuze's injunction that we ask what materials can do, which in turns, 'opens up forms of experimentation. It is a whole exploration of things, it does not have anything to do with essence.' For Lefebvre, such an approach is neither moralist, nor rationalist but 'ethological' opening up the possibility of (following Donna Haraway) a 'response-able' approach to materials. In *A Thousand*

Plateaus it is not the designer but the artisan who surrenders to the wood and follows where it leads:

> We will therefore define the artisan as one who is determined in such a way as to follow a flow of matter, a *machinic phylum*. The artisan is *the itinerant, the ambulant*. To follow the flow of matter is to itinerate, to ambulate. It is intuition in action.[45]

Importantly Deleuze and Guattari's figure of the artisan is not the same as the figure of the artisan traditional to architectural discourse who 'follows the dictates of his soul',[46] for their artisan follows the 'machinic phylum' which includes the availability of materials and resources and even the markets themselves, as well as the variegations and intensities of the materials they work.

Undoubtedly, Simondon's account of moulding brings the question of 'implicit forms' into play and there are moments in his discussion in the 'Form and Matter' chapter where he appears to valorize artisanal production as in the following description where the mould is seen as 'fixing' a set of modelling hands that would otherwise prolong the shaping 'without rupture':

> The mold limits and stabilizes rather than only imposing a form: it gives the end of the deformation and achieves it by stopping it according to a definite contour: it *modulates* the ensemble of the already formed networks: the gesture of the workman who fills the mold and compresses the clay continues the former gesture of kneading, stretching, shaping: the mold plays the part of a fixed set of modeling hands, acting like arrested forming hands. One could make a brick without a mold, with one's hands, prolonging the shaping by a fashioning that would continue it without rupture.

When Simondon claims that the mould 'limits and stabilizes' the deformations of the clay it is not equivalent to the mould 'imposing a form'. Indeed, he makes no hard distinction between forming by mould or by the artisanal process of moulding by hand. What is given through the preparations is 'the ensemble of the already formed networks' that already have the potential for the unfolding of energetic operations. What differentiates these two technical systems is how and when the dynamic operations of individuation come to a halt, so that there is no potential for further individuation.

Deleuze and Guattari take the term 'implicit forms' from Simondon.[47] He says that implicit forms are more obvious in materials which were once living – leather, bone, bark – or in particular, in a construction such as Ulysses' bed which made use of a still living olive tree, but they are no less present in matter-like materials such as clay and concrete. It is just that we need to look at a microscopic scale in order to see that clay is not 'indifferent' (in Simondon's model there can be no potential for individuation without difference, and hence unresolved tension, present within the system) but dynamic in a specific way. Clay has colloidal properties which give it a viscosity in solution with water, thus allowing it change its form, molecule by molecule in communication with the walls of the mould, and at the same time to hold together. The mould does not impose form on it but 'modulates the ensemble of [molecular] threads already formed'.[48] Concrete and clay may appear to be instances of inert homogeneous hylomorphic matter, but it is only that their implicit forms are veiled.

Simondon recognizes that there are other techniques of fabrication that are not as easily conceived in hylomorphism's terms. For example, when 'a mechanical saw cuts wood *abstractly* according to a geometric plan' it ignores the grain and torsion of the fibres in order to separate the piece into two half trunks and has no need to take implicit forms into account. Other techniques, such as splitting with a wedge, 'respect the continuity of the fibres, curving around a knot, following the heart of the tree, guided by the implicit form which the effort of the wedges reveals'.[49] Some processes, such as metalworking, use so many processes that there is no one moment which stands out as a point where matter takes form (here he does not consider the process of casting metal which is readily understood in hylomorphic terms):

> In this case, the taking of form is not accomplished in a single instant in a visible manner, but in a number of successive operations. One cannot strictly distinguish the taking of form from qualitative transformation; the forging and tempering of a piece of steel are the one anterior, the other posterior, to that which we can call the actual taking of form; forging and tempering are nevertheless the constituting of objects.[50]

Here Deleuze and Guattari seem to follow Simondon closely and take up his notion of the 'metallurgical' as that with the potential to challenge form-taking as a simple interaction between form and matter:

> [I]t is as if metal and metallurgy imposed upon and raised to consciousness something that is only hidden or buried in the other matters and operations. The difference is that elsewhere the operations occur between two thresholds, one of which constitutes the matter prepared for the operation, and the other the form to be incarnated (for example, the clay and the mold).[51]

As clay moulding was the philosopher's example par excellence for Simondon, and bronze casting for Aristotle, so too Deleuze and Guattari choose metalworking to suggest that some materials and processes provoke us to find alternatives to the hylomorphic schema for explanations of the technical operation. For other techniques, such as concrete casting, the hylomorphic schema can appear perfectly adequate as an explanation.

The found process-based description, as we have already seen in Theophilius's account of metalworking or in the specifications for making mortars, is no more likely to 'go into the mould' and give us an account of the dynamic operations individuation than any other form of description. In the examples of specifying concrete casting in rigid formwork that I will look at here, what we find described are the preliminary operations that set up the system in which the self-actualizing dynamic operations of form-taking itself arise. But, nevertheless, these techniques arise out of a detailed knowledge of processes that occur within and in relation with the mould and are developed to manage and coerce them. Moreover, many of the detailed prescriptions are designed precisely to edit out some aspects of the form-taking process at the larger scale of the finished form, as well as at the micro-scale of the concrete's 'implicit forms'. It will emerge that it is not just that the dynamic operations of form-taking are hidden from view. In fact, techniques are actively developed that erase the traces of these dynamic operations. At the level of fabrication, the appearance of the hylomorphic schema is actively reproduced.

In the rhetoric of architectural styles, the two specifications I have chosen to explore in detail describe two rather polarized approaches to working with concrete. At one end of the spectrum, in the quite typical modernist house at Egypt End that Val Harding designed with Tecton (1934–35), is the smooth white concrete of modernism.[52] In line with the then new vocabulary of the International Style, windows are large and steel framed, roofs are flat and used as terraces and the concrete wraps seamlessly around to provide a continuous enclosure punctured with openings. At the other end of the spectrum, in the concrete walls to the raised assembly hall at the Elfrida Rathbone School for the Educationally Subnormal (1961) designed by the LCC Architects with John Bancroft as project architect, is the raw *béton brut* of brutalism.[53] In this building the concrete was to be left exposed to show the marks of the timber shuttering used in its production.

Within the specifications for these two buildings process-based clauses describe the detailed ways in which concrete and formwork are prepared prior to casting for the series of operations which will occur during their encounter. The Specification for the Elfrida Rathbone School (1961) produced by the LCC represents a peak of process-based specification. Its 161 closely typed pages describe every aspect of the building process – from washing out buckets to the preparation of material samples for approval – in extraordinary detail (Fig. 4.3).[54] On the one hand, this excessive use of the process-based clause can be seen as an extension of the need to control the quality and methods of workmanship when gross tendering took off in the early part of the 19th century and the use

Formwork and moulds

C14A. For all in-situ concrete the Contractor may use either timber or steel formwork as available unless otherwise shown on the drawings or specified.

Any timbering ordered to be left in to uphold the sides of excavations or for casings, etc., will be paid for at the rate shown in the Bills of Quantities and must be measured with the Architect's representative before trenches, etc., are filled-in, otherwise no extra will be allowed in respect of same.

Formwork is to be erected true to line and to the profiles shown. Where a shuttered concrete finish is indicated the formwork shall be so designed to produce the formwork patterns shown on the drawings and shall be of rough sawn, clean new timber with a pronounced grain all to the approval of the Architect. The formwork shall be sufficiently tight to prevent any loss of liquid from the concrete and staining of the fairface brickwork and shall be of substantial construction to carry the loads due to wet concrete and any incidental loading without deformation. In all cases the Contractor shall be entirely responsible for the accuracy and efficiency of the formwork. Any fairface brickwork marked or stained shall be cut out and reinstated at the Contractor's own expense. Bolt holes will not be allowed in any finished surfaces. Unless otherwise shown the boarding for the shuttering is to be laid horizontally for beams and walls and vertically for their ends.

The formwork is to be so designed to allow of accurate adjustment by wedges or other suitable means and to allow striking to be carried out gradually without any jarring of the concrete. The erection of all formwork is to be inspected by the Architect before any concrete is placed.

Soffit boards to beams, etc., shall be placed with a camber to ensure that when the formwork is removed and the beams subjected to the full dead load the soffit shall have no apparent deflection below the horizontal.

All formwork is to be thoroughly cleaned of any old concrete and immediately before concreting it shall be thoroughly hosed down with water, holes being provided in the formwork to permit the escape of any sawdust, shavings, rubbish, etc., with the water.

Formwork is to include for all labour and materials, and for cutting to waste for the forming of all edges, projections, etc., and for all notchings at intersections, also for proppings, strutting, supporting and removal as required.

The Contractor is to allow for samples of shuttered concrete finish for the Architect's approval prior to the commencement of the concrete work generally and in the positions to be decided on site.

The following are minimum intervals of time which shall be allowed between the placing of concrete and removal of formwork:

	Cold weather (about freezing) Concrete made with ordinary cement (days)	Normal weather (about 60°F.) Concrete made with ordinary cement (days)
Beam sides, walls and columns (unloaded)	8	3
Slabs (re-propped)	10	4
Removal of props to slabs	14	10
Beams soffits (re-propped)	12	8
Removal of props to beams	28	21
	32.	

FIGURE 4.3 Specification for the Works, Elfrida Rathbone School for the Educationally Subnormal, (1961). Architect – John Bancroft with London County Council. RIBA Archives, LCC/AD/1. Source: Courtesy of the Royal Institute of British Architects Library Drawings and Archives Collection and London Metropolitan Archives, City of London.

The number of days during which the temperature has fallen below freezing point should be added to the days given in the table. For temperatures between freezing and normal these may be adjusted pro rata in each particular case.

Where sulphate resisting cement is used due care is to be exercised when striking the shuttering to ensure that the concrete has hardened sufficiently and on no account are the striking times to be less than those shown in the table above for ordinary cement.

The responsibility for the safe removal of the formwork shall rest with the Contractor, but the Architect reserves the right to delay the time of striking in the interests of the work. Any work showing signs of damage through premature loading, shock or vibration shall be entirely reconstructed at the Contractor's expense.

Care must be taken in constructing and removing formwork to give smooth edges and surfaces and true angles.

C15.   General - The concrete may be conveyed in any suitable manner from the mixer provided there is no segregation or loss of any ingredients and provided it is placed in its final position before initial setting takes place and within thirty minutes of the addition of water to the mixer and shall not be subsequently disturbed. It shall be deposited as nearly as practicable in its final position to avoid rehandling or flowing.

No concrete is to be dropped free of support from a place higher than four feet above its final position.

Every care must be taken to prevent the separation of the coarse from the finer portions of the concrete and to ensure a thoroughly homogenous mass.

If, in the opinion of the Architect, the coarse and finer portions composing the concrete are separated in any degree during the deposit, he may require the concrete to be again turned over and mixed before it is rammed.

No slopes, other than haunchings, are on any account to be allowed in connection with a deposit of concrete. Where it is necessary to form a joint in non-reinforced trench foundations the different layers are to be stepped back against an approved timber casing and at least equal to twice their depth, and with vertical faces. Before the adjoining deposit is begun, all surfaces are to be cleaned, well wetted and coated with a grout as referred to in Construction Joints (Clause C19).

Mass concrete - The concrete is to be deposited in layers from 9" to 12" thick, unless otherwise specified, which are to follow each other as quickly as possible to prevent any distinct joint between them, the whole thoroughly consolidated by working with shovels and ramming with suitable beaters. Care must be taken that the concrete is consolidated closely against the face of previously deposited concrete, against formwork, or the sides of trenches or timbered excavations as applicable in order that there shall be no subsequent settlement or disturbance of the ground.

of the process-based clause became common. Similarly, the move towards reducing the use of the process-based clause which was initiated in the 1960s and has led to its near eradication in some forms of contemporary specification, was in part a response to documents such as those produced by the LCC and the great restrictions they placed on the contractor in carrying out work in ways which were most efficient and economical for them.

On the other hand, the process-based clause offers some opportunities. Harding's house at Egypt End was built with the latest techniques of cast reinforced concrete and the specification is rich with the vivid detail of the building, preparation and striking of formwork, and the mixing, pouring, setting and finishing of the concrete (Fig. 4.4). These clauses may guide a builder through an unfamiliar process of construction and allow the architect or specifier to think through the operations of fabrication that remain outside their scope of work when clauses describe only 'work in place'. In Harding's case the intensity of detail and language suggests that this was an opportunity he relished, a means perhaps of crossing the distance between the drawn and written representations of the imagined building-to-be and the material reality of the building itself.

In both of these specifications, the processes for making ready concrete and formwork as well as a number of finishing processes after casting, seem to be designed precisely to edit out visible traces of the specific material operations of formation, even where boardmarks are marked in the finished surface of concrete. These processes seem designed to ensure that concrete appears as 'abstract matter' and formwork as 'abstract form' in a realization of the hylomorphic schema.

The specification for Valentine Harding's house shows the degree of care involved in ensuring that the cast walls conform as closely as possible to the perfect lines of the rigid formwork. The weight of wet concrete as it sets is immense and unrestrained. It could cause the formwork to bow or splay. But it is clear from the clauses of the specification that the forces of the concrete must be resisted. Clause 106 explains that any interaction between container and contained in the encounter between the two is intended to inform the finished output:

```
FORMWORK
106. Form work must be erected true to line; be properly
braced and of sufficient strength to carry the dead weight
of the concrete with any constructional loads without
excessive deflection.[55]
```

	94. The cement shall be stored in such a manner that it will be efficiently protected from moisture and the consignments can be used up in the order in which they are received.
	95. No cement which has become damaged in the storage shall be used in the work, and all lumps must be removed.
AGGREGATE.	96. The aggregate shall be composed of hard stone or ballast, free from clay, dirt or other deleterious matter. It shall pass through a $\frac{3}{4}$ inch screen and be thoroughly graded from coarse to fine.
	97. The aggregate shall not be composed of flat or flaky materials.
	98. Coal residues, such as coke breeze and clinkers, shall not be used for reinforced concrete work unless specifically approved in writing by the Architects or Engineers.
	99. The aggregates to be used are presumed to be of a nature and quality necessary for the production of concrete which will give a crushing strength of not less than 3,000 lbs at the expiration of three months. When the concrete, made from such materials, is exposed to the weather or has to retain water, the materials shall be of such a nature that the concrete will resist the passage of water. All concrete materials supplied shall be subject to the approval of the Architects or Engineers.
WATER.	100. The water shall be fresh water, clean and free from organic impurities.
PROPORTIONS.	101. That portion of the aggregate which is retained on a $\frac{1}{4}$ inch screen shall be termed "coarse aggregate" and that portion which passes a $\frac{1}{4}$ inch screen shall be termed "fine aggregate".
	102. The 1 : 2$\frac{1}{2}$: 5 concrete shall be composed of 1 part cement, 2$\frac{1}{2}$ parts sand and 5 parts coarse aggregate.
	103. The 1 : 2 : 4 concrete shall be composed of 1 part cement, 2 parts sand and 4 parts coarse aggregate.
	104. The materials shall be measured in vessels or containers of a nature which make reasonably accurate measurements possible.
	105. The Contractor may, or shall if called upon, vary the proportion of fine to coarse aggregate with a view to obtaining the densest mix, provided the amount of cement per cubic yard of concrete in position is not reduced.
FORM WORK.	106. Form work must be erected true to line; be properly braced and of sufficient strength to carry the dead weight of the concrete with any constructional loads without excessive deflection.

13.

FIGURE 4.4 Specification of Works for a House at Farnham Common, Bucks. (1934). Architect – Val Harding with Tecton. RIBA Archives, SaG/17/3. Source: Courtesy of the Royal Institute of British Architects Library Drawings and Archives Collection.

This effort to resist any deflection is in contrast with the techniques of fabric-formworking. To lesser and greater extent depending on the projects some the limiting flexible mould is allowed to bulge and splay in tandem with the forces of the liquid concrete.

As we have seen in the case of Wall One, and as we will see at Corbusier's La Tourette (and in many other examples of concrete casting in rigid as well as in fabric formwork) the surface is left rough, pitted, fissured and perhaps 'unfinished'. But at Egypt End in order to produce the desired concrete surface the shuttering must be 'perfectly smooth' and the formed concrete must itself be sanded once the shuttering is struck:

```
EXTERNAL FINISH
111. The shuttering for the external surfaces of all
walls, reveals, copings, soffits and fascias must be
perfectly smooth. As soon as the shuttering is struck and
while the concrete is still green the above mentioned
surfaces must be rubbed down with a wood float and sanded
till perfectly smooth. On no account must a cement grout
be used.[56]
```

The concrete is to remain 'pure' (any irregularities are to be polished out rather than filled with another material) but, in contrast with Wall One, it is not intended to show any of the natural variegation that might occur as it sets. The aggregate of the polished concrete will not be visible, nor will any differentiation caused by gravity in relation to the position formwork (whether for example the wet concrete pushes down on the formwork when it forms a soffit, or vertically against it in a wall). In this sense, then, both formwork and concrete are to be treated so that the concrete appears as amorphous matter which can be formed perfectly into the orthogonal shapes described by the architect's modernist concept.[57] Through the rigid formwork, form is to appear to shape matter which must in turn appear to submit entirely to its orders. Any straying from the predetermined form will be scraped away and eradicated. The extra work of polishing, sanding and bracing ensures that the casting will appear to have occurred according to the hylomorphic schema. While the same effect might have been achieved by rendering over brickwork (as indeed it was in Le Corbusier's Villa Savoye – modernsim's most famous white villa) the concrete in Val Harding's house exemplifies the dominance of matter by form.

At the Elfrida Rathbone School the formwork is also specified, but clause C14A instructs that it is to be roughly sawn and carefully designed to reproduce the patterns of timber grain that are shown on the drawings:

> C14A FORMWORK AND MOULDS
> Formwork is to be erected true to line and to the profiles shown. Where a shuttered concreted finish is indicated the formwork shall be so designed to produce the formwork patterns shown on the drawing and shall be of rough sawn, clean new timber with a pronounced grain all to the approval of the Architect . . . Boltholes will not be allowed in any finished surfaces.[58]

Although this aspect of the fabrication is registered in these concrete walls, the specifier edits others from the finished product. He insists that the grain of the shuttering is inscribed into the finished wall but traces of the boltholes must disappear. He is also concerned that the aggregate does not interfere with the fine tracery of timber and that only the finest mix is used for the outer layer:

> C10. MIXING
> Where concrete is required for exposed finished shuttered work, the first mix of each day shall consist of sand and cement without the coarse aggregate as specified and shall be spread lightly upon the bottom of the formwork in order to avoid the first mix showing excessive aggregate.[59]

By specifying the finest mix at the surface of the wall the specifier ensures that only the timber shuttering will determine the appearance of the concrete. Despite the architectural rhetoric which understands this as an 'honest' treatment of concrete because it reveals the process of production, we see that the fabrication is in fact closely controlled so that only some aspects are registered. The specification for the Elfrida Rathbone school devotes much attention to ensuring that the concrete shows no variation as can be seen in clauses C6 and C14:

> C6.   The contractor will reserve sufficient sand and gravel with Messrs. Eastwoods to complete the whole of the exposed shuttered concrete works without *undue variation of colour*.
> C14. Where concrete beams, slabs, etc. are shown on the drawings be a shuttered concrete finish, the Contractor's attention is drawn to the very high

> standard of accuracy, consistency and finish of concrete that will be required. The greatest care will be called for in formwork, mixing and placing of concrete, positioning of construction joints, removal of shuttering, etc. and the Contractor will be deemed to have allowed for this in his tender. *No rubbing down or making good will be allowed after removal of the shuttering to any of these surfaces. The resulting concrete surface is to be free of any honeycombing, cavities, pitting and any imperfections not the result of the texture of the concrete.*[60]

At the Elfrida Rathbone School the possibility that differences in the sands used to produce the concrete might be visible through some colour variation in the finished building is curtailed. Concrete's tendency to pit and fissure, due to water rising to the surface during setting or to inevitable irregularities in the sizes aggregate was considered an acceptable outcome of the casting process in Wall One. But here variation is considered to be 'imperfection' and the concrete is intended to appear homogeneous. The material specificity of concrete – both of its material parts and of the complex temporal processes of its setting – are denied here. Thus, a great deal of care is taken to ensure that the concrete appears to be a homogeneous inert matter that merely receives the imprint of the timber shuttering.

At the detailed level of these processes of fabrication the concrete at the Elfrida Rathbone School might not be so very different to the smooth concrete of the Val Harding house. In the conventions of architectural history, timber grained concrete tends to be seen as an expression of the material which is neutralized in the amorphous matter of modernism. As Alison Smithson, herself a leading proponent of brutalism, puts it, modernist buildings were 'not built of real materials at all but some sort of processed material such as Kraft cheese: we turned back to wood and concrete, glass and steel, all the materials which you can really get hold of'.[61] The process-based clauses show the extent to which, in both kinds of fabrication, the concrete's own qualities and specific processes of setting are in fact restrained and edited out. At the Elfrida Rathbone School the processes of concrete fabrication are not really revealed, rather the processes of fabrication enable the formwork to give the concrete form at two scales. As in the Val Harding house, the concrete takes its overall form at the macro-scale from the formwork, but in addition, at the micro-scale, it must take on the form of another material, the timber, which provides the limit condition to its own internal processes of slumping, and heating, and setting and drying. The concrete is mimetic as Barthes has described plastic, and it is able to imitate timber because of the fineness of its specific plastic properties. Despite the differences in the appearances of these two forms of concrete (and the rhetoric of styles surrounding them) they are both treated as matter – neutral, without identity, waiting to be given form.

Reading these prescriptions we catch sight, then, of the 'implicit forms' and energetic exchanges of both formwork and concrete in the dynamic operations of form-taking, not through their direct description as in Simondon's account of the wet clay brick, but though the techniques that ensure their effects will not be registered in the finished wall. In a letter to Derrida that was never sent, but was written in response to Derrida's proposal for the establishment of the Collège Internationale de Philosophie, and recently translated into English and published as 'On Techno-Aesthetics' Simondon makes reference to architectural fabrication.[62] He shows his appreciation of the rough concrete walls at Le Corbusier's La Tourette (1953–57) that leave visible the traces of the dynamic operations of its formation. There is, he writes, 'a politeness towards materials: we do not parge. The traces left by the formwork on the concrete of the chimney of the Dominican convent [sic] in Arbresle near Lyon are intentionally visible'. In fact, these effects are not to do with the marking of the formwork. Perhaps it takes more architectural knowledge than Simondon had to understand this. As Elisabeth Shotton has explained, the more unusual aspects of the concrete's surface, the scars of pre-stressing which appear all over it and the pitting and roughness of the casting which give the material some of its power, were in part a result of money-saving measures taken by the contractors Sud Est Travaux who took over the work midway through construction.[63]

Again, we need to be cautious in assuming that what is at stake here is the question of the imposition of form on matter, both with respect to the formwork, and to the role of prescription in the specification. In his compelling but problematic analysis of Simondon's 'Form and Matter' chapter in *Political Physics*, the philosopher John Protevi cites a passage that in some of its terms is by now familiar, although its meaning is more elusive than it first appears:

> The technical procedure which imposes a form on a passive and indeterminate substance is not just a procedure considered in an abstract way when a spectator sees what goes in and out of the workshop without knowing the process itself. It is essentially an operation commanded by the free man and carried out by the slave.[64]

Here Simondon extends the problem of hylomorphic imposition to the 'slave' who is commanded to carry out the technical operation. We can think of the builders who must follow the prescriptions of the specification when they work concrete and formwork. Protevi takes this still further and represents Simondon's free man who imposes form on passive matter, who despises 'surrender' to matter and 'only sees and commands' as an architect.[65] The architect, in imposing form, treats not only the material but, importantly, the slave who carries out his commands, as if *both* are passive and will receive his directions accordingly. In Protevi's reading a direct parallel is suggested between the architect's imposition of form on matter and their commanding of the builder. Moreover, the architect is represented as a 'free man' and the builder as a 'slave'. If we followed this logic, the clauses of the specification which specify a paradigmatic 'hylomorphic' production process such as concrete casting in rigid formwork could be seen as double impositions. They describe how form is to shape matter, and at the same time command the builder. We might then want to propose, as Protevi will, other strategies of fabrication which would 'liberate' both matter and maker. Here for example are the theorist-practitioner architects Reiser + Umemoto, describing how hylomorphism structures both the object and practice of modernist architecture:

> It is interesting that the classical form/matter duality persists in the architecture of modernism both as a fundamental philosophical concept of design, per se, and in the way design arrives in the social field. For the same duality that stipulates a hard division between sovereign form and passive matter enforces a corresponding division of labour between conception and construction.[66]

For Reiser + Umemoto this parallel between the division of form and matter and conception and construction can be undone by enabling 'virtually every facet of the design process, including the spectrum of material properties and effects' to become 'actors in this parametric field'. Although they are careful not to make overblown claims for the social effects of such altered design processes they do mobilize the rhetoric of emancipation which we have already seen creeping into DeLanda's discourse:

> While the new models of production cannot make any undue claims for their socially liberating effects, they nevertheless have increased the degrees of freedom available to the designer and, by extension, to their productions.[67]

Simondon's argument does not take this direction. He does not propose that technical operations which are represented in hylomorphic terms produce particular social relations or organizations of work and might therefore need to be challenged.[68] Instead, the argument which follows from the free man/slave passage is one about what can be expressed. If it is the case, he suggests, that the processes of form-taking are veiled in the hylomorphic schema, then they cannot be known by he who stands outside the workshop, and are not therefore 'in the order of the expressible' which can be communicated to the 'slave' who will carry out the procedure.[69] Only form and matter are visible and can be expressed. The parallel is that the worker who carries out the technical operation only sees it in terms of an imposition of form on matter as does the one who gives the command. For both the chains of processes involved remain veiled.

In this sense, then, the process-based clause in these specifications for concrete casting can be seen as more than a command imposing the architect or specifier's predetermined form on the matter (and we can be very sceptical of the extent to which they are some kind of free agent). In Simondon's sense it is also an unveiling, if indirect, of the dynamic operations of form-taking. In more recent specifications, where trouble has been taken to eradicate process-based specification in order to shift responsibility to the contractor and to give them more choice in finding economical and efficient solutions appropriate to their profit margins, these operations are once again unavailable, whether to the specifier, builder or to the researcher. Furthermore, what is revealed in the specifications we have looked at is a second veiling in the physical techniques that ensure that the full and varied effects of the processes of form-taking are edited out in the finished output. It would be a step too far to suggest that behind this 'matterization' of what is in fact a variegated and heterogeneous material is an intention to reproduce a schema that we have seen is more conceptual than actual, and realize it in the heavy, insistent and ever-present durability of concrete, but we might at least ask what drives these complex techniques.

'Rendered Plastic by Preparation': Preliminary operations

In his 'Form and Matter' chapter Simondon points out that there is a matterization of clay that is produced through the preliminary operations that make it ready for its encounter with the mould. Certain techniques applied to materials, he proposes, render them matter-like or 'plastic by preparation':

> The dominance alone of the techniques applied to materials rendered plastic by preparation can ensure to the hylomorphic schema an appearance of explanatory universality, because this plasticity suspends the action of historical singularities provided through the material.[70]

These techniques, he suggests, contribute to the apparent universality of the hylomorphic schema and they reinforce its explanatory potential because they erase the 'historical singularities' or implicit forms given through the material. The material practices through which clay or concrete are 'matterized' give validity to what is in fact a conceptual framework, rather than a physical reality. Indeed, according to Dominique Lecourt, for Gaston Bachelard (who was the teacher of Georges Canguilhem – Simondon's thesis supervisor) it is one of the characteristics of the 'new' sciences (and particularly in relation to chemistry) of the mid-20th century that they are '"artificialist", that they contain as one of their essential components a technique for the production of phenomena', that they in turn account for.[71] At least in so far as the 'objects of science' are concerned:

> Far from being poor abstractions drawn from the wealth of the concrete, [they] are the theoretically normed and materially ordered products of a labour which endows them with all the wealth of determinations of the concept.[72]

The work of conceptualization can be lent force by things and processes in themselves particularly where these are already informed by a conceptual apparatus.

The matterization of concrete, which we have seen prescribed in the process-based clauses of both of these specifications, is thus interesting in two ways. First, we catch sight of a series of techniques that render a variegated and dynamic material, such as liquid concrete with all its specific characteristics and processes 'plastic by preparation'. Through such techniques, a specific material comes to appear as homogeneous and as more readily available to the imposition of form. Constituted as such, through its preliminary operations, concrete too can take its place among bronze, wax or clay, those other plastic materials preferred by philosophers, that are themselves also prepared or matterized. Through these preliminary operations, concrete comes to appear as an instance of 'abstract matter' rather than as a material process in all its specificity, shot through with implicit forms, disparations and historical singularities arising out of actions past but nevertheless available within the material system as potential for form-taking. Matterization lends the hylomorphic schema a certain explanatory weight and doubles the 'forgetting of operation' already inherent to it, since the dynamic operations of the material are rendered invisible, and its historical singularities disappear.

Second, there is today a marked acceleration in the industrial matterization of materials. The brick industry is an ancient one, but many more materials have now been 'rendered plastic by preparation'. Glass and reconstituted stone can be cast. The building board industry uses waste products from the timber industry – lightweight timber, fibres and chips – and may mix them with resins or glues and put them under pressure – to form boards of homogeneous material which can used to replace timber or various wall finishes such as plaster.[73] MDF which was developed in the 1960s an 'all-over' material which eradicates the historical singularities of the timber from which it was made and can be moulded into any form as well as into standardized boards.[74] More recently, architect theorists and practitioners are interrogating with unsurpassed enthusiasm the possibilities of 3D printing, which dispenses with the mould as such but appears to realize a long held hylomorphic fantasy of predetermined form imposed in a material that through preparations entirely visible has been rendered homogeneous both at macro- and at micro-scales. How are we to understand this industrial proliferation of materials 'rendered plastic by preparation'? What else drives their prevalence?

Although Simondon identifies some benefits of following the grain and direction of a variegated piece of timber he recognizes that it also risks a conflict with the intended explicit form, which can be difficult for the worker. Materials carry implicit forms but they impose prior limits on the technical operation and not all tools work with them.[75] The lathe for example does not need to take the implicit forms of timber into account, and timber with a fine grain is most suited to it for it is 'almost homogeneous'. Cyril Stanley Smith has pointed out that while a craftsman can follow the 'material's local vagaries' with his tools 'the constant motion of a machine requires constant materials'.[76] Matterized materials may simply multiply the number of options for working them, reduce the number of tools needed and increase efficiency. DeLanda also refers to the increasing dependency within engineering on isotropic materials, whose properties are identical in all directions. First, he suggests that where materials are close to homogeneous their 'singularities and affects will be so simple as to seem reducible to a linear law'.[77] He recognizes that although naturally occurring materials such as metals contain impurities which make their behaviours less predictable, the industrial production of metals has led to their homogenization over the last two hundred years, not only he suggests, because of reliability and quality control but also for social reasons:

> Both human workers and the materials they used needed to be disciplined and their behaviour made predictable. Only then the full efficiencies and economies of scale of mass-produced techniques could be realised.[78]

As J. E. Gordon points out, DeLanda continues, homogeneous materials such as mild steel also make the engineer's job easier:

> At a higher mental level, the design process becomes a good deal easier and more foolproof by the use of a ductile, isotropic, and practically uniform material with which there is already a great deal of accumulated experience. The design of many components, such as gear wheels, can be reduced to a routine that can be looked up in handbooks.[79]

DeLanda stresses the industrial context of the production of materials and, recognizes, that Simondon's 'abstract matter' is materially reproduced, just as we have seen in the examples of concrete casting. Constant materials, to use Smith's wonderful term, are constructed 'abstract matter' produced in such a way that a body of knowledge can be built up about them which appears universal, and in such a way that they are suited to mass production. Without perhaps recognizing that the homogeneity of material is itself a production, Marx has drawn attention to the uniformity of silver and gold – the two materials used to make money, and makes an argument which relates to the symbolic potential of a material to embody 'abstract and therefore equal human labour':

> Only a material whose every sample possesses the same uniform quality can be an adequate form of appearance of value, that is a material embodiment of abstract and therefore equal human labour. On the other hand since the difference between the magnitudes of value is purely quantitative, the money commodity must be capable of purely quantitative differentiation, it must therefore be divisible at will, and it must also be possible to assemble it again from its component parts. Gold and silver possess these properties *by nature*.[80]

For Marx it is necessary that the materials used to make money are regular (or fungible) so that they can be easily divided up and stand in for quantities. The same volume of material, for example, will have the same weight. It is also interesting to see that Marx emphasizes uniformity as a requirement of the appearance of value. It is as if it is only by eradicating the historical singularities in the material itself and removing any traces of men's actual labour, that money can embody abstract labour. In Marx's interpretation there is more at stake in the use of a matterized material than the routinization of material production DeLanda describes. For Marx, materials that appear to reproduce the concept of matter are a precondition for the representation of abstract labour.

In these arguments 'matterization' is not only a conceptual apparatus, limited to the way we describe a process of production. It is also a material one, which exists in a particular historical, social and economic context that is currently engendering an enormous expansion in the production of constant materials. These forces must also contribute to the preliminary operations that prepare the possibility of any material production, but Simondon hardly refers to them in his accounts of technical individuation. Here for example, he extends the clay's preparation right back to its extraction at the edge of the marsh, but includes only physical processes:

> As for clay, it is also subjected to a preparation; as a raw material, it is what the shovel raises to the surface at the edge of the marsh, with roots of

rush, and gravel grains. Dried, crushed, sifted, shaped, lengthily kneaded, it becomes this homogeneous and consistent dough having a rather great plasticity to be able to embrace the contours of the mold in which one presses it, and firm enough to preserve this contour during the time necessary for that plasticity to disappear.[81]

The architectural specification is of course embedded in the industrial context of construction, but in the following example of a 'ready-made' description of the preparations for mortar in a Specification for building a workshop (circa 1833–43) from Mocatta's collection, and despite the technical language (and a different intention) there are many parallels with Simondon's description. The clause lays out in detail the processes of sourcing sand, mixing it with lime, sieving and beating it:

> One Bushel of well burnt unslacked lime made from Dorking of Mersham grey chalk or Halling Lime to Three Bushels of clean Thames sand to be taken from the river between Fulham and London Bridge. The lime and sand to be mixed together during the suspension of slacking by wetting the lime and then covering it with its due proportion of sand. The proportion of each to be duly measured. It is then to be passed thro' a fine screen of no less than 35 units at equal distance in breadth of every foot. It is then to be sufficiently beat or passed through a proper mill.[82]

Such accounts are not limited to processes carried out by hand. Although we lose the details about the provenance of materials, similar processes are described again for the mechanical mixing of concrete in the specification for the Elfrida Rathbone school:

> Mixing C10.
>
> Concrete for reinforced concrete work must be mechanically mixed in non-continuous machines. The fine and coarse aggregate and the cement shall be mixed for at least three turns in a mechanical mixer having a drum rotating about a horizontal axis, after which the required amount of water shall be added gradually while the mixer is in motion and the concrete mixed for at least two minutes until of a unified colour and consistency . . .[83]

These clauses describe getting materials ready for their use in the construction process. They are preparatory. The bonding of brick to brick is self-actualized once it has been laid on with a trowel just as concrete sets and takes form from energy already locked up in the mix. As Simondon puts it, the 'setting into form' began long before the moulding itself:

> The preparation of clay is the constituting of this state of equal distribution of the molecules, of this arrangement in chains; the setting into form is already commenced at the time when the craftsman stirs the paste before introducing it into the mould.[84]

And as with Simondon's descriptions, these particular clauses concern only physical operations. But other clauses include processes which extend beyond the physical manipulations of the material and might be called 'extra-physical'. In the Hardwick specification for mortar (1791), for example, we saw that it was to be the surveyor who would authorize the correct mix of sand and lime. In this case the preliminary operations for mortar's form-taking (as a process of bonding more than a process of taking shape) would also include getting the surveyor to turn up on site and approve the mix. Institutional operations can also establish the conditions for the mobilization or individuation of a material.

In the clauses for casting concrete, we find references to the contractor, the builder, the architect, to building standards and to procedures of measurement and control which also play a part in the concrete's preparations. For example, in clause C2 the specification for the Elfrida Rathbone School (1961) we read how bags of cement are to be checked upon arrival on site:

```
Cement C2.

Bags of cement which arrive on site must have their weight
checked, each quality of cement is to be tested on site, and
the contractor must keep enough cement on site to satisfy
the architect and send him manufacturer's advice note after
each delivery.[85]
```

And furthermore, in clause C4, how samples of each batch of cement are to be subject to quality control at the Council's testing station in special tins:

> Cement tests C4
>
> Cement will be tested at the Council's testing station. Samples to be taken from each delivery of cement as directed and conveyed by the Contractor to the County Hall . . . for testing. Tins for this purpose will be provided by the Council. Written notification of approval must be obtained.[86]

Even the workman's lunch break is incorporated into a clause concerning the care to be taking with cleaning out the concrete-mixer:

> Mixing C10
>
> On the cessation of work, including short stoppages for meals, or on any change of type of cement used in the mix, the mixers and all handling plant shall be washed out with clean water.

Preparations must also include the protection of concrete from the weather:

> Protection of concrete C22.
>
> Protect all concrete work from damage by traffic, sun's rays, drying winds, frosts running or surface water and heavy rains by covering with planks, sacking or sawdust, for a period of at least seven days after it has been laid. In hot weather the cover is to be kept well wetted.[87]

In these examples process-based descriptions drawn from practice make clear that the preliminary operations that make concrete and formwork ready for form-taking on the building site include more processes than the physical and molecular. As extra-physical processes become more and more central to the production, specification and regulation of building materials we need to extend the range of operations Simondon describes. They too need to be included and paid attention to.

That we need to take more factors than the physical in the coming into being of any technical system, is put very elegantly by Brian Massumi in a beautiful section at the start of *A User's Guide to Capitalism and Schizophrenia*. Massumi sets out a dazzling array of forces that converge in the 'wood-tool encounter' when a woodworker makes a table. They include 'a boss, a body, hands, technique, intentions, the handle of the tool' and 'a piece of wood, a customer order, rain, trucks delivery' as well as processes such as 'planing', 'the evolution of the tree's species; the natural conditions governing its individual growth, the cultural actions that brought that particular wood to the workshop for that particular purpose' and the 'particular institutionalization of craftsmanship formalizing knowledge accumulated over centuries by countless people'.[88] The breadth of his list is exemplary and better reflects the range of preparations to be found in the clauses of the specification than Simondon's near exclusive attention to the physical. But Massumi fails to ask *how* these different forces are able to interact with each other. He accepts that they are necessarily related in 'the real monism of matter' rather than ask, as we will see that Simondon does, how it is possible for previously incompatible terms to 'converge in a common operation'. And at least in the account of technical individuation it will be the preliminary operations that create the conditions for this possibility. What we have, if we stay with Simondon here, is an account of how hitherto disparate realities such as the muddy grit and reed-filled ground of the marsh and an abstract geometrical idea can be prepared towards a common operation. In fact, for Simondon, individuation or mediation is *only* possible where there is difference.

This argument becomes much clearer in 'Form and Energy', the chapter which follows 'Form and Matter' and includes his famous account of crystal formation. Here, the crystal solution, prior to the introduction of the seed crystal, is full of potential for form-taking, the conditions have been created, but as yet there is no energy exchange, no actualization of this potential, only tension and disparation. In this state it is 'metastable' – a phase prior to individuation in which there are no changes of step, when disparate aspects of a system have the potential to come into communication but as yet remain incompatible. As it turns out, the description of the clay/mould system in terms of molecular propagation allows it to resonate with the description of the crystal/solution system that it precedes. By showing how extensively the clay has been worked from extraction through to moulding Simondon establishes the build-up of potential energy in the clay that is like the potential for formation (and communication) in the crystal solution. But in the case of the wet clay brick this potential is described as having been intentionally prepared through the chains of preliminary operations. At least in the technical system preliminary operations are the processes which establish the given condition of metastability out of which individuation arises.

What I hope to offer here is a relational ontology that includes accounts of relations coming into being and, by exploring some of the preliminary operations that prepare materials for a range of mobilizations in building, to extend them well beyond the literal, physical convergences of clay and mould or concrete and formwork. In doing so we begin to catch sight of conditions that make possible relations between building materials and much wider social, economic and political realms than suggested by the form/matter schema.

5

Performance

Writing in 1989 the building technology writer Ian Chandler contrasted the 'traditional' forms of specification we have seen in Chapters 3 and 4 – naming, proprietary and process-based specification – with what he claims is a 'newer range' – the performance specification:

> Those specifications written in terms which clearly point to one construction solution, with materials named, sizes given, and perhaps even suppliers named, are within the traditional category. The newer range of specifications are based on terms which describe what the material, construction detail or component is expected to do and how it must behave during and after placement; these are categorised as *performance specifications*.[1]

The idea of specifying 'requirements which have to be satisfied' rather than prescribing materials directly had in fact been under discussion in the UK since the 1930s, and was already used as a principle in the 1940s for British Standards and Codes of Practice.[2] But in the lead up to the formation of the NBS in the 1960s and 1970s the question of to what extent performance could and should be used in specification was much debated. Only gradually did it become more widely accepted that wherever possible it is preferable to specify 'in terms of ends rather than means'.[3] By the 2000s consultants such as the UK based Schumann Smith had started developing bespoke specifications that relied heavily on performance specification. They catered initially for some of the bigger architectural practices, especially those at particular risk of being targeted by claims surveyors, who scrutinize construction information on behalf of contractors to uncover inconsistencies in the documentation. Schumann Smith designed some forms of specification to eliminate altogether the prescription of particular materials or construction processes. They avoided this 'prescriptive specification' as they termed it, and aimed instead to deploy 'open' specifying techniques that state the criteria to be met rather than the method of achieving them.

Open specification is advantageous because it shifts the responsibility for prescribing specific materials or construction solutions to the contractor, and reduces the chance of litigation against the architect. But even if this is the primary incentive, performance specification involves a great many other changes of great significance to our discussion, not least of all – as Chandler explains – that they are 'based on terms which describe what the material, construction detail or component *is expected to do and how it must behave* during and after placement'.

Unlike the category 'process-based specification', which is a term I developed based on my survey of documents for a form of description hardly identified, performance specification, is a widely recognized classification, although there are differences of definition. Stephen Emmitt and David Yeomans use the term performance to identify any alternative to prescriptive specification, and place emphasis on its contractual openness.[4] For others, such as Chandler, it is the emphasis on the *behaviour* of a material or component (as opposed to its 'species' or on how it is to be made) that is decisive. In contrast to naming or process-based specifying the material may not be named at all. Indeed, in specifying only what the material is to *do* this form of clause conceptualizes the material in a new way, through how it will act in the future, rather than as something that has been found and classified, or as something that is the result of processes of fabrication or genesis.

Taking into account this wider definition of performance specification through emphasis on how a material acts, we can identify examples of its use from outside the contemporary contractual context that gives rise to its use in recent decades. For example, at the gothic villa for Mr Usborne (1879) a timber floor is specified only in so far as it may be used for dancing:

Flooring   71.   These floors to have knots no larger than a sixpence therein . . .
Narrow floors   The drawing room floor to be partially traversed for dancing.[5]

Similarly, we can understand the clause for precast slabs in Rosenberg's sausage factory (1930s) where they are described almost entirely in terms of what they are to do as a form of performance specification:

---

20.   Fix precast slabs as shown on Drawing No.3 to
       control to the air circulation in the chambers.[6]

The concrete slab to the air raid shelter at Bedales (1940–1) is to be good enough to 'serve as flooring' with no additional finishes:

> 57. Foundations   The surface [of the floor] is to be trowel finished and suitable to serve as flooring. No screed is to be allowed for.[7]

These clauses each describe the result rather than the means of achieving it but they do not share the easily identifiable language and quantification of contemporary performance specifications. To specify a material by its properties or behaviours – acoustic insulation, heat or UV light transmission, strength in compression, traffic noise resistance and so on – the contemporary performance clause uses the clipped abstracted terminology of materials science, a discipline which emerged out of the study of polymers and was only established in the 1950s and 1960s. Properties are defined using scientific units of measurement – Megapascals (Mpa), Relative Humidity (RH), Watts per Metre squared Kelvins or u-value (W/M²K) – and stated in quantities, grades and percentages. For example, the second part of clause 215 from Mole Architects describes the performance of a suspended ceiling above a car park in the Studio extension to the school of architecture at Cambridge (2006):[8]

> K10 Plasterboard dry linings/ partitions/ ceilings
>
> 215   SUSPENDED CEILING SYSTEM
>
> TO SOFFIT OF FIRST FLOOR ABOVE CAR PARK/ WORKSHOP/ WALKWAYS
>
> NOTE. TO BE READ IN CONJUNCTION WITH SECTION K40 – SUSPENDED CEILINGS
>
> - Structural performance: The ceiling system must safely support loads including services fittings.
>   - Subject to wind/ upward pressure: No.
>   - Uniformly distributed loads (maximum): 0.6 kN/m².
>   - Additional loads/ pressures: None.
>   - Deflection (maximum) of grid between points of support: Span 400 mm. Test standard: To BS 8290-2, Appendix A.

- Fire performance:
  - Fire resistance of complete floor and ceiling assembly: To BS 476-21, 60/60160 minutes (Stability/ Integrity/ Insulation).
  - Ceiling resistance: Not required.
  - Protection to structural beams: To BS 476-23, 60 minutes.
- Airborne sound insulation performance:
  - Sound insulation of complete floor and ceiling assembly. Weighted sound reduction index, Rw (minimum) to BS EN ISO 717-1: Not applicable.
  - Other requirements: None.
- Suspension system: As recommended by the board manufacturer to complete the ceiling system and achieve specified performance.

The clauses detail structural, fire resisting and sound insulating performance – information generated and understood mostly by experts outside the architectural office. Shifts in building technology and metallurgy did not bring the materials analyst into the architectural team as Harold Rosen predicted in 1969, but it would have an enormous impact on specification practice:

> The specification writer of yesterday, with glue pot and scissors, will have to be replaced by a materials analyst of the future. The growing increase in manmade materials, the advances in metallurgy, and the advent of systems design and construction will necessitate the creation of a specialist on the design team who is versed in the science of materials, in the interrelationship of components, and in the techniques of construction.[9]

As Rosen notes, the late 1960s saw many changes take place in the production of materials and also a marked escalation in systems building. These factors contributed to making the use of performance specification a possibility at all. Performance specification requires that materials and components are subjected to tests whose results are quantifiable and recognized by the industry, and to support this manufacture needs to be large-scale and standardized. Commentators in the period leading up to the formation of the NBS recognized these changes and were behind the introduction of performance specification but in general considered its use would be partial.

From the launch of the NBS right into the first two decades of the 2000s (in its current form, as NBS Create, it is software based and less widely used outside large-scale practice[10]) the skeleton specification retained a range of forms of clause with performance clauses interspersed among them. Even

Schumann Smith's performance-oriented documents made use of other forms of clause. Indeed, both Tony Brett (of Schumann Smith) and Mark Taylor (of Allies and Morrison, who made use of Schumann Smith's services on some of their architectural projects) explained that if used on its own, performance specification in fact leaves the document too open for a contractor to price. Something more 'concrete', more *specific*, than a set of parameters and criteria needs to be provided.

But the contemporary shift towards performance, even towards what Jean-Francois Lyotard termed as early as 1979 the 'performativity principle', goes well beyond the practices of materials specification within the building industry. Outside the technical literatures I have been referring to so far in this chapter, there are also a number of discussions of performance and performativity in architecture across a wider range of practices and meanings. Their concepts of performance might just as well come from theatre studies, philosophy or feminist theory, as from engineering and management. For example, a symposium at the Bartlett School of Architecture (2013) brought theatre practitioners, architectural designers and activists, business and engineering experts together around the shared theme of 'performance'.[11] Branko Kolarevic's introduction to his edited collection *Performative Architecture*, identifies 'increasing emphasis on building *performance*' as spanning a broad range of concerns: 'multiple realms from, spatial, social and cultural to purely technical (structural, thermal, acoustical, etc.)'[12] He suggests that new digital technologies used by designers to simulate the performance of buildings are largely responsible for the shift towards performativity. Certainly, he is right that computational technologies make it possible to calculate aspects of the building's performance and incorporate it into the design process, but the requirement to predetermine performance would seem to arise from a much wider combination of factors from environmentalism to new building standards to market forces. Furthermore, as the book's subtitle *'Beyond Instrumentality'* suggests, performance-led architecture needs to be understood in relation to function but it is not equivalent to the deterministic functionalism of high modernism. According to Kolarevic performance-led architecture can also be distinguished from 'form-making', it 'places broadly-defined performance above, or on a par with, form-making'.[13]

David Leatherbarrow's essay in the same collection argues that 'the actuality of building consists largely in its acts, its *performances*' and he gives a more philosophically derived approach to the notion of performance than Kolarevic.[14] The building, he suggests should not be considered as a static object but as a constellation of actions. He extends this idea of action to materials themselves, in common with the new materialist architects and theorists we have already encountered such as Reiser + Umemoto and

DeLanda. Importantly, Leatherbarrow goes beyond the kinds of form-finding material processes that architects such as NOX have engaged with, to include the activity of building elements – from columns to cladding – and their engagement in the 'performance' of maintaining an apparently static equilibrium in the finished building:

> There is another site of architectural action in which performance is less obvious but no less determining: those parts of the building that give it its apparently static equilibrium, its structural, thermal, material stability. When discussing these elements (columns and beams, retaining walls and foundations, but also cladding and roofing systems), it is common to talk of their 'behavior' – not only talk of it but to anticipate it, even predict it.[15]

This is precisely the kind of reconceptualizing of the material elements of building that the use of the performance clause brings about at the technical level. Here however Leatherbarrow identifies only one aspect of the building's performance that appears in the performance clauses – its 'structural, thermal, material stability'.

As we saw in Section 215 for a suspended ceiling at the Cambridge School of Architecture, some clauses concerned the capacity of the building to carry its own weight, or to maintain integrity up to certain pressures of wind loading. The majority of performance clauses are indeed concerned with specifying materials and components in terms of the climatic and structural integrity of the building, but a second set of criteria not mentioned by Leatherbarrow can also be identified. These clauses anticipate particular activities and uses of the building and specify material behaviours that will provide the appropriate environment (acoustic, thermal, degree of security and so on) for them. The clause for sound insulation in the School of Architecture specification is an example of this. To determine the level of insulation the specifier must also take into consideration the activities that will take place in this environment – whether this is hearing a lecturer speak within the school, or ensuring that students do not wake neighbours with their music during an all-nighter. These clauses are particularly interesting. They extend the concerns of the specification well beyond what we have seen so far, because they are concerned with the social program of the building and its spaces, and relate to the ways the building is intended to be used. They show that material selection takes into account its relationship to an occupant, as well as its role as an element in the building as an object.

Leatherbarrow's account of performativity suggests we are free to decide whether we consider a building as a static object or in terms of its behaviours. In his view it is simply a matter of choosing one conceptualization or the other. But he also notes that 'it is common to talk of [these elements'] behaviours' as if this is a peculiarly contemporary and widely accepted way of thinking about such elements that has some historical specificity. Although tantalizing and pertinent, this suggestion is not developed in his essay.

Unlike Leatherbarrow (and most architectural and cultural theorists writing on performance) both Kolarevic and Antoine Picon in his *Digital Culture in Architecture*[16] are more inclined to reflect on the shift towards performance that is simultaneously taking place in industrial and contractual contexts, but it is the performance studies theorist Jon McKenzie in his brilliant *Perform or Else* who insists on tackling head-on the rapid extension and expansion of 'performance concepts' across domains from techno-performance to performance studies.[17] McKenzie draws heavily on Lyotard's analysis who argued that 'the imperative for performance improvement and product realization' has been driven by the desire for increased profits.[18] According to McKenzie this in turn engenders shifts towards certain frameworks of knowledge[19]:

> Performance means effectiveness, and effectiveness that, in most cases, must be quantified for measurement and endlessly qualified for evaluation.[20]

Certainly, these same factors influence the shift towards performance specification in building. But apart from telling us little about the effects of this shift (beyond the imperative to 'perform or else') and concentrating primarily on causes, both authors make rather grand claims for the power and centrality of the performance principle. For Lyotard it has come to legitimate knowledge in place of truth. For McKenzie it has even replaced discipline as the contemporary 'onto-historical formation of power and knowledge':

> Performance must be understood as an emergent stratum of power and knowledge . . . and will be to the twentieth and twenty-first centuries what discipline was to the eighteenth and nineteenth, that is, an onto-historical formation of power and knowledge.[21]

At least with respect to specification practice, I reject McKenzie's generalizing claim. Despite the fact that the escalation in use of performance specification has been very significant, we still see the persistence of variety of forms of clause in all forms of specification, even those of Schumann Smith. There is no evidence that we are moving towards a single universal form of materials description even if, as architectural practices become more and more concerned with ensuring that their specifications are claims-proof and as design and build contracts become common, specifications need to specify without being specific, and to avoid 'prescriptive' specification wherever possible. Moreover, not all materials and material processes can be performance-specified, even if new techniques have evolved that allow for an ever greater range of materials and criteria to be subject to performance specification than was initially anticipated.

Performance Specification

As early as the 1970s it was clear that performance specification demanded a large enough scale of materials production and the deployment of testing regimes that would not be possible for most building projects at the time. As Tony Allot put in one of his reports prior to the establishment of the NBS:

> The nearly pure performance specification can be appropriate, for example, where there is a large order, where design and development time can be allowed, where manufacturers or contractors have sufficient expertise, and where suitable tests exist: but the vast majority of designer specified building work is tendered for and built under conditions which make the pure performance approach difficult. Most projects are 'one off, not very large, needed quickly, built by small contractors, supervised by non-scientific staff, and do not always justify full testing procedures'.[22]

Although it seems to have been less obvious to these commentators that the form of clause used would also depend on the contractual situation, they were right that performance specification could be used only for building or materials production at a certain scale. Bespoke performance-based specifications prepared for very large projects by practices such as Fosters, Rogers and Allies and Morrison, are likely to refer to bespoke systems engineered for the particular job. In NBS documents that until very recently covered a much wider spectrum of project sizes as small as domestic refurbishments, and made use of a range of forms of clause, performance-engineered materials were still sometimes specified there by name, even when performance specification would have been just as easy.

We might also note the difficulty of submitting certain materials to performance specification if they are not consistent in every sample. Natural stone is a good example. A material such as Portland Stone varies in its physical properties from sample to sample, and from quarry to quarry. The quantification of properties in one sample cannot guarantee the behaviours of the next. So, in the following clause F21 for quarried stone wall dressings taken from the NBS we see that performance specification is avoided and it is prescribed through its source and geological description (naming):[23]

F21 NATURAL STONE ASHLAR WALLING/ DRESSINGS
To be read with Preliminaries/ General conditions.
TYPES OF WALLING/ DRESSINGS
110 ASHLAR _____ .
- Stone:
- Name (traditional): _____ .
- Petrological family: _____ .
- Colour: _____ .
- Origin: _____
- Finish: _____ .
- Supplier: _____ .
- Quality: Free from vents, cracks, fissures, discolouration, or other defects adversely affecting strength, durability or appearance. Before delivery to site, season thoroughly, dress and work in accordance with shop drawings prepared by supplier.
- Mortar: As section Z21.
- Mix: _____ .
- Sand: _____ .
- Other requirements: _____ .
- Bond: _____ .
- Joints: Flush.
- Width: _____ mm.
- Pointing: _____ .
- Features: _____ . [. . .]

But as we have already seen in the case of clay and gold, and in the transformation of variegated wood with its grain and knots to timber products such chipboard and MDF, materials can also be rendered 'constant' or homogeneous through a series of preparations. Cast or reconstituted stone is made up of cement, sand and finely crushed natural stone[24] and produces the physical consistency that performance specification demands. Immediately following the previous clause is another for stone dressings. The stone's role in the building is the same as in F21, but now that it is homogeneous cast stone it can be specified by its properties and behaviours:

> **F22** **CAST STONE ASHLAR WALLING/ DRESSINGS**
> To be read with Preliminaries/ General conditions.
> **TYPES OF WALLING/ DRESSINGS**
> 110 CAST STONE _____ .
> - Cast Stone Units:
> - Manufacturer: _____ .
> - Product Reference: _____ .
> - Absorption: As clause _____ .
> - Compressive strength: To BS 1217.
> - Cube strength:
>   Average (minimum): _____ .
>   Single (minimum): Not less than _____ .
> - Finish: _____ .
> - Colour: _____ .
> - Mortar: As section Z21.
> - Mix: _____ .
> - Sand: _____ .
> - Bond: _____ .
> - Joints: Flush.
> - Width: _____ .
> - Pointing: _____ .
> - Other requirements: _____ .

Cast stone is a material with little structural strength used in the main to resemble natural stone, so apart from the fact that it is less badly affected by frost than natural stone, its visual characteristics are most important. Here, therefore only some of the clauses for the cast stone – those relating to strength and absorption – are performance-specified. But these quantities can only be given because cast stone has already been processed in such a way that it is rendered homogeneous. No doubt it is reasonable to say that performance specification is tied to profit-driven interests of materials production at a vast scale and to increasing efficiency on the building site, and that it is contributing to transformations in the materials we use, but because not all building is at this scale, and because some materials and processes resist this form of description its use remains partial.

Early commentators also argued that not all material behaviours and characteristics could be quantified. Here for example is Carter reporting on developments in the preparation of the NBS in 1969 and responding rather disparagingly to the idea that performance might become the prevalent form of specification:

From the wilder fringes of theory the suggestion has come that the NBS should be wholly in performance terms. This is quite impracticable at the moment. For many materials and methods adequate performance criteria are very difficult to define and the tests to assess performance criteria either too difficult or too expensive to carry out. 'Non-slip-ness' of floors (if there is such a word) is one obvious example.[25]

In so far as appropriate tests have to be devised to quantify properties such as 'non-slip-ness' Carter was right but he could not have envisaged the scale of change that would in fact take place and ensure that ever greater numbers of materials and criteria could be subjected to performance testing and specification. In fact there are now quantitative assessments of the 'non-slipness' of floors and as a recent article in *NBS Journal* 14 explored, at least three different methods of testing 'slip resistant performance' in Europe have led to difficulties in specifying flooring that is manufactured outside the UK. The author Kevan Brassington (a member of the SlipSTD group which was set up to improve and standardize slip classification) explains that the UK uses a pendulum test in which a shoe-shod pendulum is swung over the dry and wettened surface to derive a 'pendulum test value' (Fig. 5.1).[26] The problem for UK specifiers is that Germany derives a different value – a 'slip-angle' or 'R' value – by using a human tester in a safety harness walking up a ramp sprayed with various solutions, which is gradually inclined (Fig. 5.2) and Italy uses a trolley-like vehicle to find the friction coefficient or 'tortus' value of the surface. Depending on the tests in use, the slip potential is given in ranges of low, moderate and high PTV or the slip-angle in degrees. Despite the fact that all methods involve quantitative abstraction the different values are not directly comparable.

There is a peculiar disjunction between the abstracted and authoritative quantities of the values derived from these tests as they appear in the scientific language of the performance specification, and the motley array of test mechanisms, which at least in the case of the ramp test pictured above rely on subjective criteria – the slip-angle is just the angle when the tester

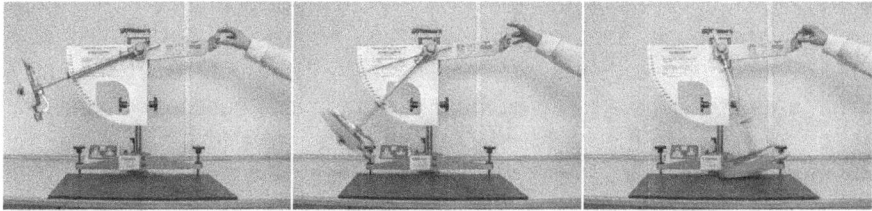

FIGURE 5.1 Pendulum test being undertaken at CERAM Research Ltd. Photographs Kevan Brassington. Source: Courtesy of NBS.

FIGURE 5.2 DIN Ramp text being undertaken at CERAM Research Ltd. from *NBS Journal* 14, May 2009. Photographs Kevan Brassington. Source: Courtesy of NBS.

'feels insecure or slips'.[27] What is clear is that despite the variation in approaches to deriving quantities for 'non-slip-ness', the drive to increasing performance specification motivates manufacturers and regulating authorities to find ways to quantify an ever greater range of properties, even those which once seemed well outside the scope of performance testing. Furthermore, the requirements involved in making performance specification possible have much in common with requirements for the establishment of standards, including the supporting framework of testing and control (Agrément Certification is a centralized system now in place for the evaluation of materials performance) and the fact that materials and products need to show no variation from sample to sample (and presumably between lab and site) if these values are to have meaning.

What we see on the one hand is the remarkable resistance of material processes to any one universal form of description in the specification, and on the other the great range of physical and extra-physical practices that make performance specification possible and their ongoing expansion. Although performance specification is characterized precisely by its focus on the future behaviours of the material, component or system, and necessarily precludes any account of the means by which those criteria are to be achieved, the range and extent of processes involved in preparing any given material for its entry into this system of description and mobilization far exceed the processes of preparing concrete for its encounter with rigid formwork. These processes may be excluded from the clause itself, but there has been a vast multiplication and expansion of the preliminary operations that make a material ready for performance specification.

In what follows I want first to propose that because it makes the *use* of the material primary (which is entirely omitted in the process-based clause, and indeed in Simondon's ontogenetic philosophy of technical systems), the performance clause establishes building materials as 'equipmental' in Heidegger's sense. The 'upfront' prescription of the material's goal is conceptually decisive for performance specification. On the one hand a goal may be ascribed simply through the act of specification itself (and the associated requisite tests and knowledge that make possible the requisite quantifications of properties), but more recently materials are now being designed and fabricated 'for a given service' both outside and within the building industry. Here, and I will make particular reference to performance-engineered glass, performance criteria determines the physical constitution of the material. The goal is physically embedded into it. Working with Andrew Barry's notion of 'informed materials', I will argue that this represents a transformation of the material that itself depends on new regimes of quantification. Not only must the physical properties of materials such as their acoustic absorption or their 'non-slip-ness' be quantified. The desired environment for inhabitants (how quiet, how easy to traverse by foot) must also be quantified so that equivalence between the two can be established. This returns us to Simondon's insistence that what is created by preliminary operations is not individuation itself, but the creation of conditions in which individuation can arise when understood as communication between orders of different magnitudes. For a material such as performance-engineered glass to be mobilized as a material that secures buildings and inhabitants against a bomb threat, for example, the conditions that put them into relation with each other must be prepared by a great range of physical and extra-physical operations that make it possible for a material's physical properties and the social and political state of security to be conceived in terms of equivalence.

'Grounded in Such Usefulness': Material as equipment

In the examples of performance specification that predate the quantification of material properties and behaviours it is immediately clear that it is what they are for – 'to be partially traversed for dancing' of 'to control the air circulation in the chambers' – that is being prescribed. But at least on first reading this, emphasis on use appears to be absent in contemporary performance specifications. Instead we see lists of properties and behaviours that may be relatively short as in the specification for cast stone wall dressings –

'Absorption: As clause _____. Compressive strength: To BS 1217.Cube strength: Average (minimum): _____. Single (minimum): Not less than ___.', or extremely long as in the following section from a specification for curtain walling in a new supermarket by Chetwoods® that uses the NBS template (the values have been taken out but were mostly 'to be confirmed by the engineer'):

H11 CURTAIN WALLING
   To be read with Preliminaries/General conditions.
DESIGN/PERFORMANCE REQUIREMENTS

305   GENERALLY:
   – Comply with CWCT 'Standard for curtain walling', Section 2 – Performance Criteria unless specified or agreed otherwise.
   – Project performance requirements specified in this subsection are to be read in conjunction with CWCT performance criteria.

311   INTEGRITY: Determine the sizes and thicknesses of glass panes and panel/facings, the sizes, types and locations of framing, fixings and supports, to ensure that the curtain walling will resist all wind loads, dead loads and design live loads, and accommodate all deflections and movements without damage.
   – Design wind pressure: _____ Pascals.
   – Permanent imposed loads: _____
   – Temporary imposed loads: _____

312   INTEGRITY: Determine the sizes and thicknesses of glass panes and panels/facings, the sizes, types and locations of framing, fixings and supports, to ensure that the curtain walling will resist all wind loads, dead loads and design live loads, and accommodate all deflections and movements without damage.
   – Calculate design wind pressure in accordance with BS 6399-2, Standard Method:
      Basic wind speed (V_b): _____
      Altitude factor (S_a): _____
      Direction factor (S_d): _____
      Seasonal factor (S_s): 1.
      Probability factor (S_p): 1.
      Terrain and building factor (S_b): _____
      Size effect factor (C_a): 1.
      External pressure coefficients (C_{pe}): _____

Internal pressure coefficients (C_{pi}): _____
- Permanent imposed loads: _____
- Temporary imposed loads: _____

313 INTEGRITY: Determine the sizes and thicknesses of glass panes and panels/facings, the sizes, types and locations of framing, fixings and supports, to ensure that the curtain walling will resist all wind loads, dead loads and design live loads, and accommodate all deflections and movements without damage.
- Calculate design wind pressure in accordance with BS6399-2.
- Permanent imposed loads: _____
- Temporary imposed loads: _____

320 DEFLECTION UNDER DEAD LOADS of framing members parallel to the curtain walling plane must not:
- Reduce glass bite to less than 75% of the design dimension.
- Reduce edge clearance to less than 3 mm between members and immediately adjacent glazing units, panel/facing units or other fixed units.
- Reduce clearance to less than 2 mm between members and movable components such as doors and windows.

330 GENERAL MOVEMENT: The curtain walling must accommodate anticipated building movements as follows:
_____

340 AIR PERMEABILITY: Permissible air leakage rates of 1.5m³/hr/m² for fixed lights and 2.0 m³/hr/lin.m for opening lights must not be exceeded when the curtain walling is subjected to a peak positive test pressure of _____ Pascals.

350 WATER PENETRATION onto internal surfaces or into cavities not designed to be wetted must not occur when the curtain walling is subjected to a peak positive test pressure of _____ Pascals.

360 WIND RESISTANCE – SERVICEABILITY: Glazed units in framing member tests must have all edges supported.

370 THERMAL PROPERTIES: The average thermal transmittance (U-value) of the curtain walling, calculated using the elemental area method, must be not more than _____ W/m²K.

380 SOLAR AND LIGHT CONTROL: Glass panes or units in curtain walling must have:
- Total solar energy transmission of not more than _____ % of normal incident solar radiation.
- Total light transmission of not less than _____ %.

385 THERMAL STRESS IN GLAZING: Glass must have an adequate resistance to thermal stress generated by orientation, shading, solar control and construction.

> 390 CONDENSATION: The psychrometric conditions under which condensation must not form on the building interior surfaces of framing members or any part of infill panels/facings are:
> – External: Summer: _____ °C maximum at _____% RH.
> Winter: _____ °C minimum at _____ % RH.
> – Internal: Summer: _____ °C at _____ % RH.
> Winter: _____ °C at _____ % RH.
>
> 410 ACOUSTIC PROPERTIES: The following minimum sound reduction indices to BS EN ISO 140-3 must be achieved:
> – Between internal and external surfaces of curtain walling: _____
> – Between adjoining floors abutting curtain walling: _____
> – Between adjoining rooms on the same floor abutting curtain walling: _____
>
> 420 FIRE RESISTANCE OF CURTAIN WALLING: To BS 476-22 and not less than[28]

Some of these properties and behaviours are given in terms of the material or system's own properties. Clause 370, for example, prescribes the average thermal transmittance or U-value that is to be achieved by the system as a whole. But the majority of the clauses here state the quantities that the materials are to withstand. So, for example in Clause 311 the system must be designed and built sufficiently to withstand wind pressures of a certain force, and both permanent and temporary loads. In themselves these quantity-based criteria to be met give little indication of why these values and not others are required, but some of the preliminary texts make very explicit what these requirements are for.

In the case of Clause 311 it is 'to ensure that the curtain walling will resist all wind loads, dead loads and design live loads, and accommodate all deflections and movements without damage'. This is the broad aim of the criteria – its 'goal' or how it must perform – but in order to reflect the situation of the curtain walling within the building or in its geographic location, further precise details are given. But whether or not the 'goal' is given directly in text, as in Clause 311, it is a necessary condition of the quantitative prescription of behaviours that what the material is to do is already known. Whether stated in the document or not, the use the material is to be put to in the building determines its description in performance terms. And at a more conceptual level we might add that use, then, becomes preparatory in the performance specification.

Importantly, Simondon excludes use as a factor in his analysis of the genesis of technical objects and systems. For him, to view technical objects for what they do is to see them only instrumentally in terms of their ends and to fail to see them in their own specificity. To define technical objects in terms of their practical uses is to miss the modifications taking place during the course of their genesis. The specificity of an object's practical ends that it is designed to meet is, he writes, 'illusory, for no fixed structure corresponds to its defined use'.[29] He continues:

> Steam-engines, petrol-engines, turbines and engines powered by springs or weights are all engines; yet, for all that, there is a more apt analogy between a spring-engine and a bow or cross-bow than between the former and a steam engine; a clock with weights has an engine analogous to a windlass, while an electric clock is analogous to a house-bell or buzzer. Usage brings together heterogeneous structures and functions in genres and species which get their meaning from the relationships between their particular functions and another function, that of the human being in action.[30]

Simondon wants rather to understand the specificity of any given technical object through its genesis, which will also enable him to explore how it 'evolves by convergence and by adaption to itself'.[31] So Simondon excludes the question of use from his enquiry in order to give an account of genesis in which the phases of the object's evolution, and not just the object's use or the designer's intention, is understood as contributing to the evolution itself in a feedback process he calls 'internal resonance'. His aim is to set up a model that can go beyond an account of the isolated technical object and better describe the complex technical ensembles which are for him changing our relation to technology. And in order to do this it is important, he writes, to understand the technical object 'as the end product of an evolution' and not as something which can 'be considered a mere utensil'.[32]

It is no coincidence here that he uses the term 'utensil'. His argument is pitched to some extent against Heidegger's discussion of tools, which makes the equipmental nature of any technology its determining characteristic. For Simondon there is more to be understood in the technical object, than its relationship to man and its usefulness:

> The thought that recognises the nature of technical reality is that which, going beyond separate objects or utensils to use Heidegger's term, discovers the essence and magnitude of technical organisation, beyond separate objects and specialist professions.[33]

Performance specification then seems particularly problematic in Simondon's terms. We might even consider it as demonstrating his argument. To describe a technical system – such as a curtain walling system for a new Sainsbury's – only in terms of the ends it is to achieve is indeed to empty out its material specificity and the processes through which it is fabricated and put into place, and to render these aspects of it invisible and as if of no significance whatsoever. The performance specification is all 'defined use' and entirely omits specific operations and energetic exchanges, despite the fact that the possibility of description in its terms (at least as constituted today) is in fact prepared, as we have started to see, through such a vast range of physical and extra-physical operations, from the grinding of stone so that it can be rendered homogeneous to the development of units of 'non-slip-ness'.

Conversely, however, to specify a material in terms of its use might be understood as exemplifying Heidegger's arguments about equipment. Curiously, despite their radical differences, the two philosophers share some common ground. First, as Muriel Combes points out in relation to the question of technique in Simondon and in Heidegger's work, both philosophers reject a means-based or instrumental understanding of technology although in very different ways.[34] Second as we will see, Heidegger is keen to avoid 'the most common way of characterizing the living being' which is 'to define it in terms of the organic as opposed to the inorganic'.[35] In this respect his approach recalls Simondon's, who is also keen to avoid any creationism or vitalism in his account of individuation. Third, Heidegger makes a claim for the preparatory role of use as characterizing equipment, as Simondon does for preliminary operations that create the conditions individuation. And fourth, Heidegger takes a critical stance with respect to the hylomorphic schema. Where Simondon sees the schema as veiling process, Heidegger sees it as failing to take account of the 'usefulness' in which the making of other things than artworks is grounded.

According to Heidegger in 'The Origin of the Work of Art' the hylomorphic schema is an account of form-making which is derived from the specific relationship of the artist to the artwork. In the production of the artwork, he writes, 'matter is the substrate and field for the artist's formative action'.[36] Just as Simondon suggests that the processes involved in form-taking are veiled from the worker who controls the production of the technical object, Heidegger makes an argument about the shortcomings of the hylomorphic schema by identifying it with a particular kind of production. When it comes to the production of equipment – such as a jug, axe or shoes – rather than the artwork, the schema is insufficient. Matter and form cannot be the 'original determinations' of a piece of equipment that is made 'for something'. In particular, it is 'both the formative act *and the choice of material*' which are 'grounded in such usefulness':[37]

> Both the formative act and the choice of material – a choice given with the act – and therewith the dominance of the conjunction of matter and form, are all grounded in such usefulness. A being that falls under that usefulness

is always the product of a process of making. It is made as a piece of equipment *for something* ... Matter and form are in no case original determinations of the thingness of the mere thing.[38]

At least in the argument in 'The Origin of the Work of Art', Heidegger draws attention to 'the choice of the material' that is already determined by the usefulness that is the determining characteristic of any piece of equipment. 'The interfusion of form and matter prevailing here', he writes, 'is, moreover, controlled beforehand by the purposes served by jug, ax, shoes'.[39] And 'it *prescribes* in each case the kind and selection of the matter – impermeable for a jug, sufficiently hard for an ax, firm yet flexible for shoes'.[40] Here, we might note, the selection of the matter is not given by a material's name, nor by how it is to be made, but in performance terms. 'Impermeable', 'sufficiently hard' and 'firm yet flexible' could each equally be stated in terms of values for absorption and air penetration, or strength, or flexibility as they are in the specification. Performance description has a particular relationship to an equipmental understanding of the technical object.

If Heidegger identifies the veiling of usefulness in part with our tendency to see all objects, even utensils, in terms of form and matter that is in fact only appropriate for the making of artworks (characterized, presumably, as for Kant, precisely by their 'purposelessness'), he also makes a second and very well-known argument that as soon as equipment is put into service or 'ready', for use its material form is 'used up' and 'disappears into usefulness'. Once formed as equipment material is merely ready to hand and unavailable to understanding:

> The production of equipment is finished when a material has been so formed as to be ready for use. For equipment to be ready means that it is dismissed beyond itself, to be used up in serviceability.[41]

In the process-based clause, as when materials are named or given through their appearance, use remains outside the description of the material. The performance clause might be seen then as drawing attention to the behaviours and capacities of materials to act, which become ready to hand or 'used up' or once building materials have been mobilized as equipment. Just as Heidegger does not name 'the kind and selection' of materials for the 'jug, ax' and 'shoes', only the way they must perform to serve the purposes of the equipment, the performance clause may well not name a material, and instead makes central the purpose it is to serve as we saw in the long performance specification for curtain walling for a supermarket. In performance clauses, even when use is not given directly, the physical specificity of a material and the processes of its coming in to being are necessarily 'used up' once it is formed and in service. Instead, it is the prescription for 'specific use' that is rendered visible in the clause, that is both determining, and in Heidegger's argument, *preparatory*.

Since we are looking at the architectural specification here it is notable that Heidegger gives a special role to 'prescription' in 'The Origin of the Work of Art'. As we will see in more detail in Chapter 6, the term 'prescriptive' is used in Schumann Smith's documents directly in contrast to the more open 'descriptive' form, to refer to the traditional form of clause that specifies the material or component by name or method of work. But for Heidegger it has a yet more nuanced meaning. It is the minimum condition that establishes something's 'equipmental character' – the 'plan' for its use:

> *A ready-made piece of equipment* is subject to some implicit or explicit *prescription* with respect to its possible use. The prescription is not given by the readiness of the equipment, but is always derived from the plan which has already determined the production of the equipment and its specific equipmental character.[42]

The defining characteristic of equipment is not that it should be used, nor even that it has the potential to be useful, but that there has been a 'plan' for it to be so. Heidegger discusses the notion of prescription of specific use and its centrality in the making ready of equipment in a section of his much earlier 1929/30 lectures, published as *The Fundamental Concepts of Metaphysics*.[43] Here it is the distinction between equipment and bodily organs, rather than the artwork, that concerns him as he sets out to explore the 'range of different kinds of beings: purely material things, equipment, instrument, apparatus, device, machine, organ, organism, animality', and asks 'how are all these to be distinguished from each other?'[44] He distinguishes between the serviceability of equipment and the 'capacity for . . .' of bodily organs[45] in order to make a distinction between inanimate objects and humans without recourse to their genesis or to some kind of vital spirit. Equipment, suggests Heidegger, can be distinguished from organs not because of its physical structure, nor because of its non-organic nature, but because it has been made 'for' serving in a particular way and that serviceability already arises out of anticipating the purpose it is intended to serve:

> Now in the production of equipment the plan is determined in advance by the serviceability of the equipment. This serviceability is regulated by anticipating what purposes the piece of equipment or indeed the machine are to serve. All equipment is what it is and the way it is only within a particular context. This context is determined by a totality of involvements in each case. Even behind the simple context of hammer and nail there lies a context of involvements which is taken into account in any plan and

which is first inaugurated by way of a certain planning . . . It is not the complexity of the structure [machine] which is decisive for the machine-like character of a piece of equipment, but rather the autonomous functioning of a structure designed for specific dynamic operations.[46]

Thus, it is certain kind of planning 'in advance' for the serviceability of the equipment that is decisive. Presumably this planning for serviceability distinguishes equipment from artworks, but it must in turn be specific to the 'particular range or possibility' for which it is intended. Equipment is not just planned for serviceability in general, bur for a specific kind of serviceability. In the specific case of a pen, for example:

Being a pen means being for writing in a particular way. The pen has been made or produced as this particular piece of equipment with this in view. The pen is finished and ready only when it has acquired this particular serviceability, this particular range or possibility, in the course of its production.[47]

Heidegger recognizes that equipment may only 'offer the *possibility* of serving for' what has been prescribed. Prescription is a condition of being equipment. A specific use must have been intended in the plan, but it is not necessary that the specific use is in fact fully determining:

Such readiness can also imply the process of making something ready in and through production and preparation, a preparation or making-ready which procures and produces the ready-made product as something independently present at hand and present to hand for use. Equipment offers the possibility of serving for . . . , it always has a particular readiness for . . . which is grounded in the way it has been made ready.[48]

Here, in a manner that recalls the clauses of the architectural specification, Heidegger leaves two blank spaces (. . .) in his sentence that can be filled in by some specific use or another. In doing so he tries to demonstrate this notion of prescription in the text – it must be there but the specifics of its content are irrelevant.[49]

Prescription, then, for Heidegger or the plan for usefulness, rather than usefulness per se, is what ontologically grounds equipment and distinguishes from artworks in 'The Origin of the Work of Art' essay, and from bodily organs in *The Fundamental Concepts of Metaphysics*. In common with Simondon's preliminary operations prescription is preparatory, but it only has this grounding role with respect to equipment. Equipmentality, or being for specific use, is excluded from Simondon's analysis. His preparatory operations create instead the conditions for form-taking or individuation. Heidegger and Simondon offer us very different, even contradictory, frameworks for conceiving of technical objects and materials. Given that the process-based clause in the architectural specification conceives of materials in the terms of Simondon's operations-oriented framework and that, in contrast, performance specification conceives of them in Heidegger's equipmental terms, we might also say that these two forms of clause make use of distinct and mutually exclusive conceptions of materials, both of which are distinct from hylomorphic conceptions of matter. This is of course not to suggest that specifiers choose between one form of clause or another out of allegiance to one philosophical framework or another. And of course we have seen that in fact a number of forms of clause may sit side by side in any one document. Rather these forms of clause emerge out of particular industrial, statutory and contractual conditions. They relate to the different ways that materials are prepared for use in building whether to meet specific criteria or make in-situ casting possible, but at the same time they are conceptually distinct from each other and from the notions of matter that architectural discourse habitually deploys for considering building materials.

With respect to the performance specification, we can say first that it conceives of the material in terms of its use. But as in Heidegger's equipmental prescription it is the very act of describing a material in performance terms, or planning for it specific use that renders it equipmental even if the content is left blank. The specific use need not predetermine in any way how the material is used or what it is like (its 'readiness') only its status as equipment. But on the matter of distinctions between one material form and another, or the specifics of the material processes that prepare a material for specific use Heidegger's account has nothing to say. In what follows I turn to performance-engineered materials, such as the types of glass produced by Pilkington, which are becoming increasingly common. These materials are not just specified for a specific use, they are *designed* to meet the criteria for specific use. For these materials it is more than a matter of prescription, or even making prescription possible. The range of preparatory operations – both physical and extra-physical – is vast, and furthermore they create the conditions in which the goals that materials are prepared for can become social as well as physical.

'For a Given Service': 'New glass performances'

In their book *Glass in Building* that was written in part as promotional material for the glass manufacturer Pilkington and their products, David Button and Brian Pye make the claim that new kinds of glass technologies are making it possible for the materials to perform in new ways:

Today, window glass performs many functions additional to its fundamental daylighting role. The advent of new requirements for building such as the accommodation of information technology, and the development of new glass technologies to meet those requirements, for example the deposition of electroconducting coatings, are generating *new glass performances*.[50]

Similarly, the Pilkington website also categorized their glass products not by name but by their performances, or as they put it: 'by benefit'– solar control, thermal insulation, fire protection, noise control, safety/security and self-cleaning. If so far we have looked at performance specification as a particular form of material description that, albeit, requires a vast institutional, technical and contractual apparatus to render it possible, we might also note that performance specification is especially suited and made use of to specify performance-engineered materials such as these kinds of glass that are designed and manufactured to meet the requirements of specific use. Furthermore, these uses exceed the structural and constructional performances that maintain a building's stability and integrity that we saw earlier were identified by Leatherbarrow in his account of a building's performance, and they may also be 'new', insofar as they extend the range of functions that building materials are designed and intended to provide. These 'new glass performances' include for example, the capacity of a material assemblage to withstand military or terrorist attacks ('safety/security') or transform what was previously the cultural imperative to keep a building clean through human labour into behaviours that are now embedded as a property of the material itself ('self-cleaning').

According to Michelle Addington and Daniel Schodek the development and take-up of performance-engineered materials has been slower in the building industry than in other industries, but nevertheless it represents a significant shift away from traditional materials, in so far as they are selected or engineered 'to meet a specifically defined need':

> For many centuries one had to accept and work with the properties of a standard material such as wood or stone, designing to accommodate the materials' limitations, whereas during the 20th century one could begin to select or engineer the properties of a high performance material to meet a specifically defined need.[51]

Moreover, they continue, those needs must themselves be defined in terms of 'an array of physical behaviours' that already represent a move away from a conventional architectural understanding of materials.[52] I will argue, furthermore, that the design of materials towards 'a specifically defined need' often involves the embedding of functions that would previously have been carried out by people into the physical constitution of the material, or what Bruno Latour has called 'delegation' from humans to non-humans that necessarily involves 'a change in the very matter of expression'.[53] Functions that were once visibly orchestrated, contestable and available to change become fixed, invisible, and non-negotiable.

To get a sense of this shift we can compare two German parliament buildings of the 20th century, both of which made use of glass screens between the visiting public and the debating chamber to manifest an ideological commitment to transparency in democracy in their political architecture. According to Peter Buchanan the equating of an open political process with visual transparency and glass walls is a particularly German concern. It became explicit during the postwar reconstruction of Germany and can be traced further back to Hannes Meyer and Hans Wittwer's entry in the 1927 League of Nations competition where Meyer's design report apparently stated there would be 'no back corridors for backstairs diplomacy, but open glazed rooms for public negotiation by honest men'.[54]

Hans Schwippert's self-effacing modernist Bundeshaus opened on the banks of the river Rhine in Bonn in 1949. In her book on transparency in German political buildings Deborah Ascher Barnstone includes a marvellous photograph of members of the public standing on tiered benches that were provided in the courtyard adjacent to the parliamentary chamber. They peer in through the glazed wall at the debate inside. One window is open, and some members of the public appear to have climbed into the chamber to listen. But, Barnstone tells us, this possibility was limited only to two lengths of wall and in any case it was soon curtailed due to security concerns. Before long there were drapes across the windows, and access to the courtyard had been restricted to parliamentarians and their visitors.[55] In this building, the glass barrier gives visual access, drapes control light and sound transmission, and rules, fences and guards provide the security function.

Underneath the famous glass dome at Norman Foster's renovated Reichstag in Berlin, opened in 1999, there are two further domed glass screens between the public and the debating chamber with a press gallery between. From within the glass dome, it is almost possible to look down through the shallower glass domes on to the heads of MPs in the chamber below, and perhaps, with a pair of binoculars to decipher the comments they make in notebooks on their lap. But if we consider the glass screen at the Reichstag in terms of what it *does* – its performances – a set of conflicting concerns that give it specific characteristics beyond its visual transparency are revealed; the glass has been engineered to withstand attack and to achieve the acoustic separation that in fact prevents participation in the debates below. As Hisham Elkadi has observed, 'Visitors to the Reichstag . . . cannot listen or be listened to and are denied any real interaction . . . in contrast to . . . the public gallery in the House of Commons in London.'⁵⁶ At the visual level the glass suggests openness and accessibility, but at the level of what the glass *does* it hinders access and defends against it as a threat. Moreover, where at the Bonn Bundeshaus the guards and drapes that performed these operations were highly visible and could be manipulated, at the Reichstag the operations are physically embedded in the glass itself and invisible.

It could be argued that the glass screens of both buildings can be described in performance terms. The large windows at the Bundeshaus can certainly be considered in terms of 'what they do', but they predate the testing and manufacturing conditions that would enable the glass to be performance-specified. Glass in traditional specifications may be specified with a proprietary name such as 'Crown Ratcliff Glass' as in the Specification for the Sessions House at the Old Bailey, or perhaps in terms of standards 'best seconds' and 'good thirds' as in the following clause from the 'Glazier' section of a specification for the Bristol Asylum for the Blind and adjoining chapel published in 1835:

> The windows on basement floor, to the men's work-room, and basket-shop, to be glazed with good thirds. The chapel and vestry windows, and the whole of remaining windows of house, to be glazed with the best seconds, free from colour and air bubbles . . . the whole of the windows throughout the buildings to be left perfectly clean and whole at the finish of the works.⁵⁷

Although the evenness and cleanliness of the glass alluded to here might contribute to its capacity for transparency and light transmission, the specification describes the glass as a particular material to standards common to the building industry, rather than through what it should achieve. Moreover, many of the functions that are now embedded into performance-engineered glass are carried out by other means at the Bundeshaus, from the drawing of drapes to the policing of the inner courtyard.

For legal reasons Fosters' could not give me access to the documents for the Reichstag glass screen, but they at least confirmed that the glazing was laminated with a PVB interlayer.⁵⁸ On rare occasions, even when the full apparatus for performance specification is in place, glass may be specified without reference to what it is to do, as in the following 'traditional' prescriptive form of clause for DSDHA's Pottersfield Park Pavillion (2006) that uses the NBS format:

> 07 Glass:
> - See clause 40/02
> - All glass to be double glazed and both panes laminated
>   Glass to be free from scratches, bubbles, cracks, rippling, dimples and other blemishes / defects. Heat toughened glass must be subjected to a heat soaking regime (8hrs) to minimise the risk from nickel slufide inclusions. All panes must be heat soaked. Provide certified evidence of treatment and all panes to be permanently marked accordingly.
>
> 08 Laminated glass:
> - See clause 40/02
>   Clear float glass with an interlayer of polyvinylbutyral (PVB), not less tan 0.375mm thick, or methyl methacrylate resin.
> - Final selection of glass and glass thicknesses to be approved by CA, prior to ordering materials.
> - Seal edges of laminated glass with materials compatible with the interlayer. Any delamination will be rejected.
> - Edges of laminated glass to be finished with no lipping between panes.
> - Notwithstanding any other requirement the lower pane of all horizontal / slope glazing is to be laminated.[59]

But it is much more likely that performance specification was used at the Reichstag, and that the glazing system would have made use of glass that was designed for its capacity to meet acoustic and security criteria. In what follows I draw on another specification for curtain walling by Schumann Smith, the company who produced many of Foster and Partners' bespoke specifications at the time, identify different categories of performances that can now be attributed to materials.

Typically, in technical literature and in most specifications with the exception of the NBS, different types of performance clause are divided into broader categories which suggest important distinctions between uses. In *Material Architecture: Emergent materials* John Fernandez suggests there are two basic types of performance requirement. 'Essentially', he writes, 'the vast majority of performance requirements can be divided between load transfer and barrier system requirements'.[60] Schumann Smith's Specification for The H. . .y Building (2006) attempts to bring all performance clauses into a single section[61] and also lists them under separate categories of requirement type: 'General', 'Structural Performance', 'Environmental Performance', 'Acoustic', 'Fire' and 'Durability'.[62] 'Load transfer requirements' relate to the structural integrity of the materials and components and are included under the category 'Structural Performance'. In some of these clauses the loads the material or component are those of the building's inhabitants, as for example in the clause below setting out the strength of doors:

H11.1411 **Strength of Doors**

a) Ensure that the doors including ironmongery, meet the 'heavy duty' category as defined in DD 171 or an equivalent international standard. At the same time doors shall comply with and not compromise the other sealed performance criteria for the works.

b) Provide evidence to demonstrate that the doors, including ironmongery, have been tested to meet the minimum acceptance criteria given in DD 171 for the following:
  i) Slamming shut impact
  ii) Slamming open impact
  iii) Heavy body impact
  iv) Hard body impact
  v) Torsion
  vi) Download deformation
  vii) Closure against obstruction
  viii) Resistance to jarring and vibration
  ix) Abusive forces on handles[63]

'Barrier system requirements' include weatherproofing and resistance to other kinds of impacts, or the constructional integrity of materials and systems. In specifications for curtain walling most clauses relate to barrier system requirements since it is primarily acting as a barrier and must withstand likely levels of moisture, wind and high and low temperatures to maintain its constructional integrity. The Specification for The H. . .y Building (2006) also groups these clauses together in one section 'Environmental Performance' (which includes, for example, 'H11.1412 Moisture Movement' and 'H11.1418 Weather and Water Penetration Resistance). It is within 'barrier requirements' that we might also identify a further category of performance requirement concerning the comfort and security of the building's inhabitants, such as acoustic, thermal and lighting control, which are intended to orchestrate their environment. In fact the majority of performance requirements Fernandez identifies concern the involvements between different materials in maintaining the internal equilibrium of the building and in relation to its environment, such as clause 370 'thermal properties' which concerns the extent to which the glazing system will maintain an internal temperature and clause 410 'acoustic properties' which is to do with the sound reduction to the internal spaces in the Specification for a supermarket (2005).

This third type of clause, which prescribes how materials are to perform to provide the acoustic, thermal and visual environment of the user, is particularly interesting and significant. These clauses are concerned more with the effect of the materials on the experience of the inhabitant than with the integrity of the building and to some extent they anticipate the behaviours of the inhabitants too – the degree of heat conservation specified for a space might be affected by whether the inhabitants are anticipated to be sitting (and not contributing much heat themselves) or doing something more strenuous. Although performance clauses always describe materials through physical properties of materials, the specific uses to which these refer show that the 'serviceability' of the building is prescribed in ways that extend to social parameters. We might suggest that these kinds of clauses contribute a 'socialization' of materials that was not possible in the same way before their implementation.

What is particularly interesting in more recent performance-led specifications is that these occupant-based specifications are the ones that remain the responsibility of the architect. The following 'performance matrix' has been developed by the in-house specification team at Allies and Morrison, a large-scale and well-respected commercial practice who specialize in high quality office buildings with glass systems engineered façades, where the system design is contracted out to specialist façade engineers and fabricators. The matrix brings together all the performance clauses for glass rain screen cladding to avoid repeating the list in different locations throughout the specification, each time such a system is specified for walls, doors or windows.[64] Along the top axis are glass rain screen cladding 'systems' 'S1' 'S2' and so on, used in different parts of the building, for example to the reception area, to an office, a storage area, for glass doors etc. Some values will be null values; for example, there is no need to worry about sound reduction to a storage area, and air leakage requirement is not an issue if the system is a door.

What is noticeable about this the performance criteria listed in this table, however, is that none of them relate to the structural or constructional integrity of the rain screen cladding system. This has all been given over to the façade engineers. All that remains for the architect, who knows how the parts of the building are intended to be used, is to specify the environmental performance: based on what kinds of accidents (whether impacts from 'hard bodies' such as stones, to 'soft bodies' such as footballs) might occur in each part of the building, what activities will take place in particular spaces, what items of value will be stored where and so on.

Clause no.	Performance criteria	S1	S2
	Test impact		
	Hard body		
	Requirement		
	Soft body requirement		
	Air leakage requirement		
	Water penetration requirement		
	Minimum light transmission %		
	Maximum G value %		
	Thermal transmission Vision W/m²K		
	Thermal transmission Opaque W/m²K		
	External sound reduction dB R_{TRA}		
	Flanking sound reduction Vertical		
	Flanking sound reduction Horizontal		
	Security requirement		

The specific uses prescribed here concern how the fabric of the building is to behave to suit or even circumscribe the intended activities of inhabitants (mostly humans but not in all cases if this was a cow shed, a shark tank at an aquarium, a kidney bean warehouse, a bank safe and so on). These requirements are carefully specified and planned, and may also be the subject of building regulations yet they are rarely considered as one of the ways that space is controlled despite the fact that they may have significant impact.

Furthermore, it tends to be behaviours of this sort that are now engineered into materials. The online catalogue from Pilkington demonstrated the extent to which they have multiplied beyond the traditional daylighting role. If in the Bundeshaus windows most of these performances would have been achieved through other more apparent means – brushes, cloths and graft keeping windows clean and rules, and regulations, fences and guard patrols providing security – a variety of technologies make it possible now to embed these behaviours into the glass itself. Toughened glass provides additional strength, and various kinds of laminated glasses, including those with high performance plastic interlayers (as nearly as possible transparent of course) prevent glass from shattering into dangerous shards. Self-cleaning glass works by using a microscopic photocatalytic layer which makes use of light to break down organic matter and another 'hydrophilic' layer which prevents water from forming droplets so that it comes off instead in sheets taking the organic matter with it. A whole range of fine coatings are now applied to glass to reduce reflectivity, solar gain and reduce heat transmission. Hans Joachim

Gläser, dates the introduction of glass coatings as early as the 1950s (when they were used in the panoramic windows of German sightseeing trains).[65] Schwippert's 1949 Bundeshaus predates their appearance and relied instead on drapes. Today the use of these almost invisible films in glass is widespread and as we see in the illustration below from a *Glass in Building*, these films are defined not in terms of what they are but what they are *for* (Fig. 5.3).

It is notable that Addington and Schodek recognize the impact of these shifts on the industry – both at the level of what such materials can do and in the ways they are described, even conceived.[66] These shifts are yet more marked in the 'smart materials' that are the subject of their book, and are responsive and react dynamically to environmental changes. Commentators from other disciplines have also recognized the significance of the development of performance-engineered materials and in many cases they emphasize their design for a specific use. Cyril Stanley Smith, for example, predicts that materials engineering 'for a given service' will not only put the engineer at the heart of 'most new projects' but that these will include projects that are 'social in nature' and involve the bringing together 'of fields that because of their special complexities have been unrelated: it would minimize the difference between the scientist and those who try to understand the human experience'.[67] In their *History of Chemistry* Bernadette Bensaude-Vincent and

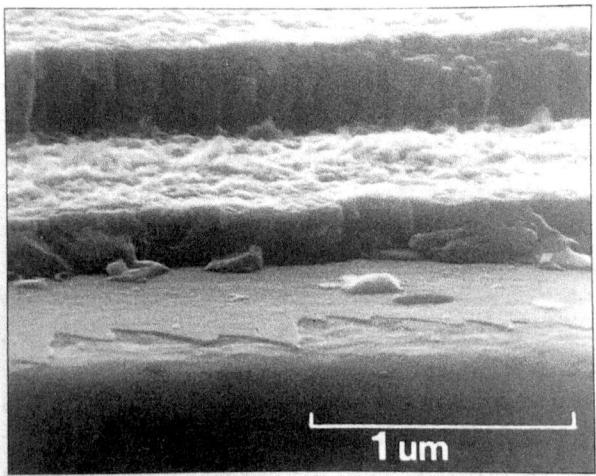

Figure 11.6
Three-layer coating on glass,
Pilkington Technology Centre,
Lathom, UK

The layers are
(1) thickness 200 Å, for adhesion;
(2) thickness 2250Å, for emissivity;
(3) thickness 2400Å, for reflectivity.

FIGURE 5.3 Illustration of Pilkington glass coatings described 'for' their performances – image taken from David Button and Brian Pye (eds) *Glass in Building* (Oxford: Butterworth Architecture, 1993). Source: Courtesy of Nippon Sheet Glass Co., Ltd.

Isabelle Stengers suggest with respect to the history of chemistry, that a significant change takes place in the 1970s when 'chemistry was put to work on a made-to-order-civilisation':

> In the 1970s . . . there arose a new orientation toward research on specific characteristics and specific materials. Plastic materials were diversified, geared to specifically targeted performance, and designed as a function of the final product. This new trend manifested itself not only in plastics but in materials as a whole. After having achieved mass production and standardisation, chemistry was put to work on a made-to-order civilisation.[68]

They use a range of terms to describe these new engineered materials – 'materials a la carte', 'materials on demand', 'made-to-order materials' and 'functional'. Like Smith, who proposes that materials engineering is a science of aggregates, rather than molecules or particle physics, they explain that these are usually composite materials and able, importantly, to perform in a number of ways or to integrate 'the maximum number of functions into a material'.[69] In the range of glasses that Pilkington produce, it is perfectly possible to include films and coatings that enable multiple functions and we might also note that some actions that glass is now able to perform, such as self-cleaning are themselves new. But what is decisive for these commentators is that the materials are designed with 'specifically targeted performance' as an aim. There is a shift here from 'prescription' in Heidegger's sense, where equipment is planned for a specific use (albeit with a vast industrial apparatus of testing, material homogenization and regulation in the case of performance specification), to materials in which specific use is inscribed at a physical level.

Latour offers a number of accounts of the 'inscription' of specific uses or 'goals' into the technical objects and systems we make. Of course for Latour, both humans and non-humans act (and increasingly in hybrid networks that are human and non-human) or perform, but design often involves the transfer of action from human to non-human in what he describes as a 'translation' of a 'script from one repertoire to a more durable one'. 'The direction' he writes, 'goes from a provisional, less reliable one to a longer-lasting, more faithful one'.[70] A typical example of his is the door closer as a permanent and entirely reliable 'translation' of the uniformed doorman. The door closer (or 'groom') can be 'polite' and give you plenty of time to walk through, or 'rude' requiring a hefty shove to open the door, and then giving you only a few moments before the door slams hard on you. Also typical, is his marvellous account of the speed bump, where the goal – for drivers to slow down on campus – is delegated from policeman or sign at the side of the road into concrete:

Instead of signs and warnings, the campus engineers have used concrete and pavement. In this context the notion of detour, of translation, should be modified to absorb, not only (as with previous examples) a shift in the definition of goals and functions, but also *a change in the very matter of expression*. The engineers' program of action, 'make drivers slow down on campus,' is now articulated with concrete. What would the right word be to account for this articulation? I could have said 'objectified' or 'reified' of 'realised' or 'materialised' or 'engraved' but these words imply an all powerful human agent imposing his will on shapeless matter, while nonhumans also act, displace goals, and contribute to their definition . . . I want to propose yet another term, *delegation*.[71]

We might understand some of the functions that glass performs in terms of this delegation from human to non-human. Cleaning shifts from the hard brush, soapy water and graft of an employee, into an action that the surface of the glass sets in motion when it rains. The security function of glass – its capacity to withstand bullets as seen in the table below – might equally be performed by some vigilant guards at the Reichstag. But in general, the actions that buildings and their materials provide are usually much more distributed than in the examples Latour chooses. Goals are more likely to be translated from one material actor to another, rather than from humans as Latour's already anthropomorphic term 'delegation' suggests. Despite his claim that 'You discriminate between the human and the inhuman. I do not hold this bias but see only actors – some human, some non-human, some skilled, some unskilled – that exchange their properties'[72] Latour tends to see a direction in inscription that travels from humans to non-humans, presumably because he wants to persuade social constructionists quite directly that objects and actor-networks have a crucial role in how we are '"allowed", "permitted", "enabled", "authorized", to do things'.[73] In the case of performance-engineered glass such goals might be tested for and given in very explicit terms. Here, for example, is a table produced by Pilkington and included in *Glass in Building*. It gives the thickness required for their glass products in relation to specific guns:[74]

Composite structure required for different levels of attack

	Parabellum 9mm	Magnum revolver 44	Nato rifle 7.62mm	Shotgun 12 bore
Laminated glass (mm)	20–40	40–60	50–80	30–50
Glass/ polycarbonate composite (mm)	15–25	25–30	30–50	–

Through testing and recording the strength of two kinds of glass – the traditional laminated glass and the performance-engineered glass/polycarbonate composite – to withstand particular impacts, it becomes possible for the glass specifier to select the correct thickness based on the perceived likelihood of each of these specific attacks (or on the thickness they can afford!). The specification of the glass does not only depend on a stand-alone property of the gun. The capacity specified is to withstand a particular named act whose likelihood is determined by a specific defined social milieu. In Latour's example of the National Rifle Association's defence that it is the gun holder who shoots not the gun he extends his argument that both humans and non-humans are agents to suggest more provocatively that they can then also both have goals.[75] In this sense the 'goal' of these Pilkington security glasses, is shown to be to withstand attack from specific gun shots, and it is this goal which is materially inscribed through the thickness of the laminate or composite used. Thus a goal (to protect people from gunfire) is carried into or 'delegated' to the material register that could be achieved in other non-physical ways – for example by legislating against gun ownership.

In addition, Latour understands that the inscription of goals is only possible in the context of other work setting the conditions that enable it to be effective. For this Latour has a second term – 'pre-inscription' – which refers to:

> All the work that has to be done upstream of the scene and all the things assimilated by an actor (human or non-human) before coming to the scene as a user or an author [by which he means designer or engineer][76]

'Pre-inscription' is here the preparations that make it possible for goals to be translated from one repertoire to another. It recalls of course the preliminary operations of Simondon, and also the vast industrial apparatus of testing, material homogenization and regulation that is required for performance specification. To support performance-engineered materials the scale and complexity of this apparatus expands yet more. We have seen already that there are tests and quantifications for the non-slip-ness of floors, or for the capacity of glass to withstand the impacts of bullets from specific guns. To specify each of the performance criteria listed in the Allies and Morrison performance matrix requires another set of relevant tests. In order, for example, to quantify the capacity to withstand different kinds of 'bodies' (from 'soft' footballs to 'hard' stones) at different sizes and speeds some impacts can be directly tested. To test bullet-proof glass, for example, the UK glass manufacturer Romag offers their own 'in-house testing facilities, which include a high velocity cannon for firing projectiles to test the impact resistance of glass and a ballistic firing range'.[77] In other cases, such as the need to test for the impact of human

FIGURE 5.4 Test rig. Figure 16.3, Test rig, Pilkington Glass Ltd., St Helens, UK, in Button and Pye (eds) *Glass in Building* (Oxford: Butterworth Architecture, 1993), 258. Source: Courtesy of Nippon Sheet Glass Co., Ltd.

body (a 'semi-hard' body) these tests are simulated. Shown is a rig which uses a head-shaped leather punch bag filled with lead shot to test the impact of the human (semi-hard) body on a plane of glass (Fig. 5.4).[78]

It is not enough then, to simply quantify the properties of the material itself. What is decisive is its behaviours in relation to other (specific) factors that if directly tested can be named (as with the guns), but if indirectly tested must be given as quantities in order that equivalence can be established. As early as 1969, some years before performance specification became mainstream in the UK, P. J. Sereda pointed out in an article for *Canadian Building Digest* entitled 'Performance of Building Materials' that the prediction of performance depends on more than the properties of the material to be used. The context in which a material is to be used, will also bring it into contact with specific environmental conditions which must themselves be quantified and tested:

> Exact prediction of performance requires a complete understanding of material properties, the processes involved in the interaction of the material with its environment, and the environmental factors to which it will be subjected. The only complete test of performance is trial by use. Prediction will always be limited by lack of complete knowledge; trial by use will be limited by time and inability to extrapolate to new conditions. Test methods can be used to supplement knowledge and experience in predicting performance.
>
> Some test methods depend on empirically derived relations between observed behaviour and some easily measured physical property; others subject the material to environmental conditions simulating those to be expected in practice.[79]

Sereda is mostly concerned with the durability of materials, and the environmental factors he discusses tend to be climatic. But his argument gains yet more significance where the environmental factors that are taken into account involve the comfort of a building's inhabitants. The inscription of these goals into materials such as performance-engineered glass have become increasingly prevalent in the intervening years. For example, where heat transmission is the criteria for the use of a piece of glass or glazing system something must also be known about the expected warmth of the space it is to contain, and this involves precise quantifications of inhabitants' expected behaviours, including for example the metabolic heat generated by various activities (unsurprisingly the guidance works with the bodily surface area of 'a typical man'):

Activities	W/m²
Reading	55
Typing	65
Walking	100
Lifting	120

The specification of one of these quantities then assumes and reproduces the appropriate conditions for the activities that are to occur in any given space.

Additionally, according to Pilkington, these figures need to be moderated by the degree of insulation of clothing that the reader or typist or walker is expected to be wearing, given in 'clo', 'a relative value, with 1.00 being the measure for normal indoor clothing *for a man*'.

Clothing	Clo
Underwear, socks, shoes	0.70
Underwear, shirt, trousers, jacket, socks, shoes	1.00
Underwear, shirt, trousers, jacket, heavy quilted jacket and overalls, socks, shoes	1.85

Here, once again, quantification enables direct relationships to be established between what clothing someone is wearing and a material component of the space that will enclose them. Once these tests and equivalences have been made between two very different modalities, it becomes possible for assumptions about the building's inhabitants and the activities they are expected to be carrying out are able to directly inform the selection or design of a material. Without these chains of techniques such a connection could not be made. Glass could be named, its properties could even be given, but it could not be described and mobilized in terms of a specifically defined need.

We might then recognize that the testing and quantification of some of the goals or specific uses which enable glass to be performance-specified are concerned with putting the material to social use, and that these criteria become embedded into the selection process. In a particularly revealing passage Pilkington promotes the performance-tested and designed glass they produce by explaining that a good level of thermal comfort can attract employees:

> Demands for improved comfort by the building occupants will increase in industrialised counties during the 1990s for several reasons. First, the older working population will have higher comfort expectations. Second, competition for employees will generate incentives by employers; a healthy comfortable working environment will be just such an incentive in the United States where the number of young adults will fall by 40% creating a shortage of entry level employees.[80]

In this appeal to the corporate client, they make a direct link between the material and a good workforce. Such a link can only be demonstrable because a relation has already been made, through testing and quantification, and in particular through the making equivalent of the comfort of inhabitants and the physical properties of the materials they manufacture.

With respect to security glass, the assessment that predetermines the material build-up of a glass screen in a politically important and therefore vulnerable building such as the German parliament must depend on an even wider set of social and political factors. As the façade engineer Sergio de Gaetano explains in his principles for designing 'enhanced façades' a 'threat assessment' is 'the starting point of any bomb blast evaluation' and is a 'risk assessment of the likelihood of an attack on a building'. Furthermore the factors to be taken into account include 'the value of the building, its function and the nature of business', the building's location, accessibility and proximity to other possible targets, as well as 'the political stability and history of terrorism in the city and in the country':[81] These factors are a crucial issue is clear from the example of Richard Rogers and Partners' Welsh Assembly Building (2003) where another glass screen gives visual access to the debating chamber. Albena Yaneva has discussed the use of the glass in this building to represent democracy, and points out that transparency is only one property of glass. 'Even reinforced glass that is very difficult to break', she writes, 'cannot easily escape the connotation of fragility. Which property of glass should *represent* democracy?'[82] And indeed, according to the specifier for the project there was a tension between the desire for visual transparency and the requirement for its security function. Midway during the detailed design phase the September 11th attack on the World Trade Centre took place and the subsequent rise in the security requirement made a full redesign necessary.[83] As Yaneva suggests, asking 'how can the political be made with glass?' may not be answered by asking about the symbolism of the material. There is more at stake here than the purely functional requirement that these factors are to be taken into consideration during design. Once performance-engineered materials are being used, and a chain of processes are implemented that makes it possible to make equivalences between the wider social climate and the physical composition of a material such as glass, these factors become goals for the materials design that are literally embedded into the physical fabric of the building and reproduced in their performances.

Madeleine Akrich has suggested that when assumptions such as these are involved in the designs of objects, they 'prescribe' them back to humans, and Latour has proposed (giving objects language in his typically anthropocentric fashion) that to some extent this might be considered as 'the moral and ethical dimension of mechanisms':

> I will call, after Madeleine Akrich, the behaviour imposed back onto the human by nonhuman delegates *prescription*. How can these prescriptions be brought out? By replacing them by strings of sentences (usually in the imperative) that are uttered (silently and continuously) by the mechanisms for the benefit of those who are mechanized: do this, do that, behave this

way, don't go that way . . . As Akrich notes, prescription is the moral and ethical dimension of mechanisms.[84]

It is true to say that expectations about use influence the selection of the glass in very precisely calculated and articulated ways and that the materials are in Akrich's sense 'prescriptive' in as much as they are guided by these assumptions. And although social goals are inscribed into performance-engineered glass, it is not possible to point to one or more goals or human actions a particular material is replacing, as it is in Latour's examples of the door closer or speed bump. In this sense the goals inscribed into a building material have become more generalized and more difficult to discern than in other technical objects. Their concealment may be further enhanced by their literal invisibility, as in glass screens at the Reichstag where films and interlayers were designed precisely to be as far as possible transparent. For Akrich the translation of social relations into technical objects naturalizes the social processes through which they are built up:

> It makes sense to say that technical objects have political strength. They may change social relations, but they also stabilise, naturalise, depoliticise, and translate these into other media. After the event, the processes involved in building up technical objects are concealed. The causal links they established are naturalized. There was, or so it seems, never any possibility that it could have been otherwise.[85]

Performance-engineered glass can be seen then as a means of choreographing social relations that both conceals and naturalizes its 'political strength'.

What is decisive about performance specification is that it conceives of materials in terms of specific use (or as 'equipmental' in Heidegger's sense) and we have seen that in the context of the building industry that this depends on more than a thought experiment. A vast apparatus of preparatory processes at industrial and institutional levels are needed to render performance specification possible and meaningful. Moreover, the rise of performance specification has taken place in tandem with changes in contractual arrangements and in particular with the development of new kinds of materials designed for specific use where goals are embedded in the physical constitution of materials. These goals are increasingly social, even political, in nature. They take us well beyond the physical operations that are considered and prescribed within traditional process-based specification, but for the most part, as Akrich argues, they are stabilized, naturalized and depoliticized in their translations into other media, so that we fail to see and interrogate them. The decisions involved in their making, calculations, tests, manufacture and regulations take place for the most part outside the sphere of architectural practice, where they are all too easily passed over or ignored, especially when materials are seen as little more than the available matter for form-making. We might, by way of a first conclusion, and following Latour and Akrich, at least hold open that regimes of performance description and engineering are rendering building materials more 'matters of concern' than 'matters of fact'.[86]

We have seen so far that an equipmental conception of materials is precisely one that Simondon avoids in order to examine the genesis and evolution of technical systems beyond the design intentions and plans for their use. He insists that while some technical objects remain 'abstract' and are simply realizations of the functions they are intended to perform, their realization in the world also starts to suggest new potentials beyond the initial diagram in a process he calls 'concretization'. To some extent we might also consider Latour's notion of 'translation' in this sense. In his account of the speed bump, he argues that because of the translation into another material register, it is not possible to think of it as simple mediator or conduit for the goal, or in the hylomorphic terms of the imposition of will (that would be the social constructionist argument). For Latour the material intervenes in unforeseen ways, independent of the engineers and designers' intentions and provides its own impetus to the development of further objects and systems and might also produce new goals. The speed bump, for example, no longer calls on drivers to consider pedestrians on campus and slow down. Instead, it appeals to their desire to protect the chassis of their car. In this way Latour makes central the goals or 'specific uses' that Heidegger sets out and at the same time acknowledges the contribution that the translation into the material makes, that is so important for Simondon in his accounts of the evolution of technical objects and systems. Moreover, Latour draws attention to the array of forces that come together in the realization of the speed bump and asks us to consider how, through the speed bump there is a 'mediation . . . in which society and matter exchange properties':

> The speed bump is ultimately not made of matter; it is full of engineers and chancellors and lawmakers, comingling their will and their story lines with those of gravel, concrete and paint, and standard calculations. The mediation, the technical translation that I am trying to understand resides in the blindspot in which society and matter exchange properties.[87]

What may be most interesting here, is precisely *how* it becomes possible for 'engineers, chancellors and lawmakers' to comingle their will 'with gravel concrete and paint, and standard calculations'. What preparations render such an 'exchange' possible, and indeed what may be its effects? In our example of performance-engineered materials and specification we have seen some of the processes that render equivalences between social goals (among others) and the micro-constitution of performance-engineered glass and make possible such a mediation. These chains of preparatory processes can also be considered in terms of Simondon's preliminary operations. We will remember that for Simondon preliminary operations only prepare the possibility of individuation in a technical system. It is not the individuation itself that is created but (as we will see in Chapter 6) the conditions for individuation, understood as the dynamic operations of disparate magnitudes now in communication. If the preparations of clay and mould establish the possibility of their communication in form-taking, then we might also understand testing, quantification, matrix-making, specification writing and so on as the chains of processes that create the conditions in which mediation between such disparate magnitudes as terrorist threat and the micro-constitution of glass is possible.

6

Systems of Material

Tempting as it may be to argue that in the context of the building industry performance specification superseded naming and process-based description, or even (with Jon McKenzie) that it should be understood as one instance of a much wider shift to the performance paradigm as 'emergent stratum of power and knowledge', the evidence is to the contrary. Specifications from the 19th and 20th centuries, and even the NBS of the 21st tend to include a variety of forms of clause, often to be found side by side in a single document. As we will see in what follows, even in the performance-oriented 'open' specifications of Schumann Smith there are moments where the injunction to avoid closed specification breaks down or produces strange new forms of clause. And because performance specification demands particular industrial formations of materials testing, quantification and production, it excludes some kinds of materials and forms of construction that must therefore be defined through alternative means if they are to be used.

I propose in this chapter that the variety of forms of clause uncovered through this research is in fact evidence of a variety of *systems* of material, in the sense of what Simondon calls the 'complete system' of individuation. As Miguel de Beistegui has written, Simondon 'prefers to speak of "systems" rather than "substances".'[1] His system is organized around individuation (and constituted then by a kind of interior tension or potential for compatibility rather than by external limits). It comprises two main phases – the metastable state which is rife with potential for resolution, and individuation itself in which the resolution between at least two previously incompatible orders of magnitude unfolds and 'accomplishes itself by itself'. These aspects are important because, first, in the case of Simondon's technical system they include preparations in the definition. It is not a case of positing a substance that can be prepared this way or that with different outcomes. Rather it is a case of those preparations altering altogether the system itself – such that in the case of building materials, a different form of description is required for each system of material (and not merely for each substance, or historical era, or contractual situation). Second, in Simondon's technical system it is not the individual that is created, but the conditions for individuation or mediation. This is the most surprising and speculative result of pursuing his account of individuation in relation to building materials and the forms of clause. What is prepared is not the material itself, but the conditions for its possible mobilization that we can think of in terms of a communication between orders of magnitude.

Simondon's complete system was already introduced here in the discussion of preliminary and dynamic operations in Chapter 4. He first refers to it in the Form and Matter chapter:

> It is the *system* constituted by the mould and the pressed clay which is the condition of the taking of form; it is the clay which takes form according to the mould, not the worker who gives it form. The man who works prepares the mediation, but he does not accomplish it, it is the mediation which accomplishes itself by itself after the conditions have been created.[2]

The system here refers both to the preliminary operations which prepare the clay and mould for form-taking, and to the dynamic operations of form-taking itself. Importantly, the worker only *creates the conditions* for form-taking, which is rendered here as a *mediation*. In the case of the process-based clause, it is detailed descriptions for such preliminary operations that are given; the mediation (whether setting hard to bond bricks in the case of mortar, or concrete organizing itself in tandem with the formwork) or individuation 'accomplishes itself by itself after the conditions have been created'. In the case of performance specification any account of the preliminary operations is excluded from the scope of the description. Instead, as we have seen in Chapter 5, a series of preparatory practices and techniques make its peculiar abstracted language possible. All manner of production techniques, from 'matterization' to micro-engineering, all kinds of tests – often surprisingly rudimentary and crude – create the conditions through which material behaviours can be directly quantified in relation to social and environmental factors. This is a very different kind of mediation to that of material taking form. It is interesting to note, also, that in the case of the process-based clause the effects of the system of material do not extend to the end user of the building. They are limited to the material effects of the dynamic system within the mould, and in some rare cases to its effects on the fabricator (for example where they are asked to show judgement about the workability or stiffness of a mix), whereas in some forms of performance clause the effects on the end user become the central organizing principle.

Adopting Simondon's complete system for a theory of materials is decisive in two key ways. The first is the centrality it gives to 'active mediation' that in Simondon's account allows two realities or different orders of magnitude such as the clay and the mould to come into communication, and itself constitutes individuation. This has led me to what may be the most surprising consequence of following Simondon's model, and to the most significant contribution of this book towards a theory of building materials – that if we think in terms of systems of materials, it will be the relations each system makes possible that is decisive rather than any physical or aesthetic characteristic of the material itself. The second key characteristic is that Simondon's 'complete system' is discrete:

> The initiative of the genesis of substance returns neither to the raw material as passive nor to the form as pure: it is the *complete system* that generates, and it generates because it is a system of actualization of potential energy, joining together in an active mediation two realities, of different orders of magnitude, in an intermediate order.[3]

There is not one general field of potential out of which any individuation can arise. Rather, Simondon argues, there are within each system, different preindividual fields that have the potential for different kinds of individuation. This is true at the general level of modes of individuation identified by Simondon – technical, physical, living and psychic – but also for specific systems within each of these modes. Within the physical mode, for example, Simondon includes both the motion of a pendulum (in which energy transforms from potential energy when the weight changes direction, to kinetic energy during the swing) and crystallization (in which the potential energy of the crystal solution is transformed to structure in the crystal). That for Simondon systems are discrete and multiple, allows us to understand what I will call the 'persistence of variety' of forms of clause in the specification. These are evidence of multiple systems of material, that exist alongside each other on the building site and in the specification, rather than aberrations, or traces of obsolete forms that have not quite been edited out of the documents. The system of material is characterized by its preliminary operations, including both physical and extra-physical preparations, *and* the specific kinds of relation or compatibility (mediation) they render possible.

Proposing first that Andrew Barry's development of the notion of the 'informed material' is useful to a theory of materials that seeks to recognize both physical and extra-physical preparations, and is helpfully understood in relation to Simondon's notion of information as individuation, this chapter goes on to set out Simondon's 'complete system' and make a case for a theory of systems of material in which forms of specification would themselves figure as an aspect of the preliminary operations making materials ready for building. Finally, taking contrasting examples from contemporary architectural practice, I show that these extra-physical preparations have effects not only on the kinds of materials that can be employed in building, but also on what architectural design can and cannot be concerned with. The effects and implications of systems of material may be more far reaching than the use of physical substances in building.

Informed Materials

In an intriguing but undeveloped passage in Bernadette Bensaude-Vincent and Isabelle Stengers' *History of Chemistry* the authors distinguish between traditional materials and a newer kind designed with a given service in mind. Their distinction recalls architectural technologist Ian Chandler's account of performance specification. But Bensaude-Vincent and Stengers go a step further to argue that the new kinds of materials now (like performance-engineered glass) in fact 'embody a different notion of matter' to that of the hylomorphic model:

> Whether functional or structural, new materials are no longer intended to replace traditional materials. They are made to solve specific problems, and for this reason they embody a different notion of matter. Instead of imposing a shape on the mass of material, one develops an 'informed material' in the sense that the material structure becomes richer and richer in information.[4]

This new notion, they suggest, is of an 'informed material'. What do they mean by this? And if this shift is primarily a change in the degree to which materials are informed, then why do they suggest that nevertheless it constitutes a *rupture* with the traditional notion of matter? Are they proposing there is more than one notion of matter? With respect to the variety of forms of clause we have seen so far – naming, process-based, performance – that each conceive materials in very different, even contradictory ways, could we also say that each embodies a different notion of matter?

The idea of the 'informed material' is not returned to in their book, nor is a source given for the phrase although it appears tantalizingly in quotation marks, but it is examined by Andrew Barry in a brilliant article entitled 'Pharmaceutical Matters; The Invention of Informed Materials'. Taking as his case study the chemistry of new drugs emerging out of contemporary pharmaceutical research and development, he asks 'How are we to make sense of the idea that materials can somehow become "informed" or as they suggest, "richer and richer" in information?'

In Barry's account the molecules produced by contemporary research-driven pharmaceutical companies can be described as informed because they are not simply 'bare molecules'. As I have argued with respect to building materials and their extra-physical constitution, and particularly in the case of performance-engineered materials, Barry identifies a wide range of factors in addition to the purely physical at work in the production of new pharmaceutical compounds:

> Pharmaceutical companies do not produce bare molecules – structures of carbon, hydrogen, oxygen and other elements – isolated from their environments. Rather they produce a multitude of informed molecules, including multiple information and material forms of the same molecule . . . The molecules produced by a pharmaceutical company are already part of a rich informational material environment, even before they are consumed. This environment includes, for example, data about potency, metabolism and toxicity and information regarding the intellectual property rights associated with different molecules.[5]

The practices of chemistry are of particular interest to Barry because chemists must always understand their molecules in terms of the environment they are designed to operate within, and this will almost certainly be different to the conditions of the lab in which they are produced:

> For chemists, the fact that molecules have changing properties depending on their associations is an everyday reality. The molecule that is isolated and purified in the laboratory will not have the same properties as it has in the field, the city street or the body.[6]

In present conditions this environment is not merely climatic or biological – and in part Barry's insight is to show that it also comprises legal, economic and commercial data. Like Stengers and Bensaude-Vincent, Barry draws on the work of Alfred North Whitehead (mostly on a small section of *Process and Reality*) for his discussion. The concern for Whitehead is largely physical science's tendency to treat entities such as molecules as if they can be abstracted from their environments, and from their histories:

> The atomic material entities which are considered in physical science are merely these individual enduring entities, conceived in abstraction from everything except what concerns their mutual interplay in determining each other's historical routes of life-history. Such entities are partially *formed* by the inheritance of aspects from their own past. But they are also partially formed by the aspects of other events *forming* their environments.[7]

Although Whitehead's point here demands we think historically even about physical entities, it applies as equally to molecules formed when brewing a herbal remedy on the hearth as to molecules formed in a shiny corporate lab in Switzerland. For Whitehead we must consider the necessary role of specific environments in the forming of all physical entities. In this sense, they are all 'informed'. This might help us to account for the role of the informational context in forming Barry's pharmaceutical molecules, or indeed for how data about risk or standards of comfort might inform performance-engineered glass. Certainly, these environments may have become, as Barry puts it, 'increasingly dense, spatially extended and ... formed not just through laboratory syntheses and tests, but through virtual libraries, computational models and databases'.[8] And certainly, pharmaceutical R&D 'makes the informational content in invented materials more clearly visible',[9] as do the discourses and practices around Pilkington's performance-engineered products. But in the terms of Whitehead's model, these tendencies would be changes in degree more than changes in kind. We can understand the environment of pharmaceuticals research as ever more saturated with information, and the materials it produces as therefore 'richer and richer in information'. However, this change in degree amounts neither to the shift away from traditional materials that Addington and Schodek recognize as brought about by materials engineered 'to meet a specifically defined need' nor to the rupture with traditional notions of matter that Stengers and Bensaude-Vincent suggest, only to an intensification. Nevertheless, there seems to be something to these claims, and as we have seen, that it may in fact amount to a change in kind is also reflected in the very different form the performance specification takes to the process-based clause, and the contrasting concepts of matter they depend upon and invoke.

Barry's account of Whitehead's argument emphasizes that the environment is not simply external to the object. Rather, he writes, 'An environment of information and material entities *enters into* the constitution of an entity such as a molecule'.[10] While for Whitehead all entities are (and have always been) informed by all other entities since they are part of what he calls an 'extensive continuum', we might recall that in Simondon's account of individuation there is no relation prior to individuation itself. There may be a state of metastable equilibrium in which there is the potential for mediation, as yet unresolved, but this is not prior to or outside the particular system of individuation. This preindividual (or metastable) phase of a system is specific to that system, rather than belonging to a general condition of all there is or 'extensive continuum'. In this phase the possibility of mediation or 'information' is latent and internal to each system of individuation. As Adrian Mackenzie has put it:

> Information, at least as Simondon understands it, occurs whenever a transductive event establishes an intermediate level at which disparate realities can be articulated together. Information in this sense does not arrive

from outside a system and need not be a discrete event. It eventuates whenever the 'unresolved incompatibility of a system becomes an organizing dimension in the resolution' of that incompatibility.[11]

Information, in Simondon's sense, 'does not arrive from outside a system and need not be a discrete event'. And information is, as Simondon puts it, 'identical to the fact that something individuates'.[12]

Simondon's account of the 'complete system' of individuation and his emphasis on the central role of mediation (as information) can help us out of the impasse that has appeared here in trying to understand what Stengers and Bensaude-Vincent might mean by a 'new notion of matter'. Rather than suggest that entities form through ever shifting associations in a continuum, Simondon's model makes the disparations necessary to any mediation both central *and specific* to any given system of individuation. Mediation is only a possibility within what is already constituted as a system, and the processes of communication are particular to that system. As we have seen, what creates the conditions for communication or individuation, at least as described in the case of technical systems, are a series of preliminary operations. They prepare the metastable condition in which disparation is latent but not yet unleashed.

I propose, then, that what is being engineered or prepared in the case of performance-engineered materials, or indeed in the case of Barry's pharmaceutical molecules, is the potential for new kinds of mediations in Simondon's sense. It becomes possible to make a molecule that is also at the same time necessarily connected to patent law, or a type of glass that has the demonstrable capacity to be mobilized to suit a particular degree of terrorist alert. These relations were not always possible, nor has something changed that renders them universally possible today. They belong to specific systems of material, and their possibility has been prepared through specific chains of processes, such as the tests and quantifications that we have seen are involved in the constitution of performance-engineered glass. If we think of building materials not as physical entities, but as systems of material, then we might be closer to understanding how new systems of material can emerge, that involve radically new kinds of relation. We might start to see how a variety of systems of material could exist side by side in building practice, each requiring radically different kinds of descriptions that suggest a range of 'notions' of material. Although each system of material would still be understood in the general terms of Simondon's model of individuation, the kinds of mediation or information central to each could be radically different from one system to the next. In comparison to the more traditional relations involved in the clay/mould or concrete/formwork system (whose preparations are evidenced in process-based specification), the relations between social and molecular realms made possible in new drugs and building materials designed for a given service (as evidenced in performance specification) can certainly be said to be 'new'.

Simondon's 'Complete System'

The clearest account of the two phases of individuation that comprise Simondon's 'complete system' of individuation is given in his discussion of crystallization in the 'Form and Energy' chapter that directly follows on from 'Form and Matter'. Simondon is particularly interested in how the crystal seed entering a supersaturated crystal solution begins the formation of the crystal. He describes the solution as in 'metastable equilibrium', a state which in scientific terms implies a system that is neither stable or unstable, but rather on the point of changing state if there is some kind of very slight interaction, such as when a sound or slight motion can bring about an avalanche, or the seed enters the crystal solution. Simondon's footnotes in *L'individu* often have the function of moving the argument back and forth between chapters, and connecting concepts between the disparate examples he uses. So, in the 'Form and Matter' chapter, before his has introduced either the process of crystallization or the notion of the seed, a footnote explains first that in the case of the clay/mould system the seed is the mould, and gives two further examples of natural systems which form due to a seed:

> The mediating singularity is the mould here; in other cases, in nature, it can be the stone which starts the dune, the gravel which is the germ [seed] of an island in a drifting river of alluvia: it is of the intermediate level between inter-elementary dimensions and infra-elementary dimensions.[13]

Simondon raises a number of issues in relation to the seed and the amorphous crystal solution. First, he notes a change of state from isotropic to geometric, which is not in any sense ordered by the seed crystal, only set in motion by it and depends on the potential energy already available in the metastable equilibrium:

> The initial structure of the seed can't entail positively the crystallization of an amorphous body [corps] unless the latter is already in a metastable equilibrium: there must be a certain energy in the amorphous substance [substance] which receives the seed crystal, but as soon as the seed is present, it possesses the value of a principle.[14]

Although the seed has no ordering potential to organize the solution, it appears to have the value of a principle. But this is deceptive. Despite the fact that the seed comes from outside it (or 'enters in' to recall Barry's phrase) it will be incorporated into the structure of the crystal once individuation commences. In this sense it becomes literally internal – not external – to the crystal. We can't 'speak therefore of an energy exterior to the crystal, because this energy is carried by a substance which is incorporated by the crystal in its own growth'.[15] This is an important distinction between crystallization and the clay/mould system where the mould, as seed or 'mediating singularity', is not incorporated. The notion of incorporated seed will enable Simondon to build his account of the individuation of the vital organism, and to set up an alternative to the notion of the gene as a kind of informational blueprint for the genesis of an individual.[16] As we will see later the seed is also internal in the sense that is internal to the individuating system comprised of both preindividual and individuating phases.

In each of these cases, then, what is crucial for Simondon is that the seed is not an organizing principle, and also that it is internal to the individuating system. But second, it must also be the case that the seed has some kind of identity with the metastable equilibrium, just as the metastable equilibrium must be able to receive it:

> Furthermore, the interiority of the structure of the crystal germ isn't absolute, and does not govern in an autonomous manner the structuration of the amorphous mass; in order for this modulating action to exert itself, it is necessary that the structural germ brings a corresponding structure to the crystalline system in which the amorphous substance can crystallize; it is not necessary that the crystal germ is of the same chemical nature as the amorphous crystalline substance, but there must be identity of the two crystalline systems, in order for the harnessing of potential energy contained in the amorphous substance to be able to operate.[17]

It is this compatibility between seed and solution, the potential for corresponding structuration latent within each, that Simondon calls 'information'. Although he takes the term from cybernetics he goes to great lengths to distinguish his notion from what he sees as a prevalent misreading of it – that it arises from outside a system and has an organizing force – in the same way that form can be misconceived if it is understood in the terms of the hylomorphic schema. Toscano summarizes his approach as follows:

> Simondon's approach to information is twofold. On the one hand, he applies to information theory and cybernetics the same critical parameters that lay the groundwork for an operational ontology of individuation; on the other, he presents a reformed concept of information as the key to a philosophy that would finally give the preindividual its due. That the modern

concept of information is here subjected to critique should come as no surprise. We could even say that in its most 'dogmatic' uses, whether in philosophy or science, it is the bearer of a grand synthesis of the three main principles of individuation that come under Simondon's attack: as unit-measure which atomistically composes organization and quantifies degrees of order, it mimics atomism; as an expression of the unilateral relation between model and copy, it reinstates the Platonic archetype; finally, as a source of organization which is separate from matter or 'substrate independent', it is the contemporary heir to Aristotelian hylemorphism. In order to counter the widespread tendency to consider information as the principle of individuation that can synthesize all others, Simondon is obliged to rescue it from hypostasis and track its specifically operational reality.[18]

Tracking the 'operational reality' of the crystal as it forms, involves recognizing this informational compatibility (that is at once necessarily disparation without which there would only be a single term) at first between solution and seed, and then between the solution and the exterior surface of the crystal once it has begun to grow. We might also recall here Simondon's detailed account of the ways in which both mould and clay were made ready for the possibility of their coming together in form-taking. Here the possibility of their compatibility or communication had to be produced through a long chain of operations. This is a rather different notion of information to Barry's. Rather than consider information entering into or saturating the pharmaceutical molecule, we would need to think of it as what renders any possibility of relation between say, intellectual property rights and the physical molecule. In the following passage this is described as 'a certain potential' of the metastable equilibrium for the crystal seed to be 'meaningful' in as much as it can bring about release of the potential in individuation. In this sense information is a potential itself already within the system:

> Information must be understood in the real conditions of its genesis, which are the very conditions of individuation in which it plays a role. Information is a certain aspect of individuation; it demands that before it there is a certain potential, in order for it to be understood as having a direction (without which it is not be information but only weak energy). The fact that information is genuinely information is identical to the fact that something individuates.[19]

In the case of Barry's pharmaceutical molecule this potential for a certain compatibility would have been prepared by a chain of operations, much as techniques of quantification and testing prepared the possibility of equivalence between the performance of a sheet of glass and any given climate of security and threat in a particular location.

For Simondon individuation is always the 'joining together in an active mediation two realities, of different orders of magnitude'. But he nowhere states conclusively what kinds of realities this might refer to or gives a clear definition of the differences coming into compatibility that might count as 'information' as far as he is concerned. Instead, we have to infer this from the kinds of examples he introduces to support his arguments. In *L'individu* these often involve energetic exchanges, as in the case of crystallization, which furnishes him with an example of an exchange from the energetic state of the solution to the structure of the crystal. But more often it is a matter of communication between very different scales, described in terms of differences in orders of magnitude or 'grandeur' to cite the French term he uses. For example, in a footnote to a section on the living being, for whom information is interior (rather than on the surface as it is in the crystal) he describes photosynthesis as the capacity of the living being 'to bring different orders or magnitude into relation with one another: that of the cosmic level (as in the luminous energy of the sun, for example) with that of the intermolecular level'.[20] Here, photosynthesis is a process that establishes (or finds) the latent compatibility of the sun and the biochemical structure of a leaf, whose difference is not so much in terms of their energetic state but in terms of their very disparate magnitudes. We might also recall that in one of his more suggestive identifications of possible orders, Simondon describes the mould as comprised of inter-elementary chains that are 'vast sets locking up the future individual (working, workshop, press, clay)'. Here communication is not limited to being between two orders, and the range of orders of magnitude are greatly expanded to the social and institutional.

Simondon uses the language of 'ensembles' and 'sous-ensembles' (given in Taylor Adkins' earlier translation as 'sets' and 'subsets') to describe these relations in a system of individuation. Systemic relations are only possible where there is information, or the potential to become compatible. Without information there are only parts of a set or 'ensemble' and not a system at all. Simple heterogeneity is not enough to establish metastability as Simondon explains in *Individuation psychique et collective*:

> Simple heterogeneity without potentials cannot instigate a becoming. Granite consists of heterogeneous elements (quartz, feldspar, mica), and nevertheless it is not metastable.[21]

No mediation can arise between granite's parts. We saw already that this would also be the case if we poured sand into the mould, rather than the prepared clay. In the same set of footnotes to Chapter 1 of *Individuation psychique et collective* Simondon gives the potential for mediation as:

> The essential distinction between an *ensemble*, the unity of which is structural and not energetic, and a *system*, a metastable unity made of a plurality of ensembles between which there exists a relation of analogy and an energetic potential. The ensemble does not possess information. Its becoming can only be that of a degradation, an augmentation of entropy. On the contrary, the system perseveres in its metastable being due to the activity of information which characterizes its systemic state.[22]

Metastability is then, the particular state in which mediation between disparate orders can arise, and it is information and the potential for mediation which characterizes the system, with its specific but undetermined capacity for resolution in individuation (and for further individuations in some cases, particularly for living systems) that constitutes the system.

So far, we have looked at this notion of the creation of compatibility between previously disparate orders, as it appears in his Simondon's accounts of individuation, but the idea also has a grounding role in his elaboration of 'invention' in a course he delivered between 1965 and 1966, entitled 'Imagination et Invention' in the edited collection *L'Invention Dans les Techniques: Cours et conferences*. Here he tracks a vast range of technical objects from levers and winches to techniques of photography and agriculture. For Simondon 'invention' distinguishes the 'technical object' from the 'created object' in so far as the technical object is the resolution of a problem that necessarily exceeds the terms in which the problem is set. So for example, in the case of the winch or hoist, a single operator is able, gesture by gesture, 'to act as if he displaces a tiny fraction of the load, compatible with his strength'.[23] The mediating hoist has the effect of dividing an indivisible load into weights that can be lifted by the operator. And without it, the operator could not have any effect on the load at all. There would be no possibility of communication between them. So in this sense then, invention is the resolution of the operator's aim to lift the load, and at the same time it is also 'the discovery of mediation between the two orders'.

Invention must necessarily involve mediation, but it can also arise without any intervening mechanism. Plants, animals, and indeed humans may provide the resolution of a problem. He launches his lectures with the beautiful example of a road that is blocked by a heavy boulder. After some time, enough travellers have been halted in their journeys by the obstacle, that when they all push together they can move the boulder. It is their combined effort that transforms each individual into a force that can act on a part of the boulder. They become, writes Simondon, 'sous-ensembles' (or 'subsets'), organized according to mode of compatibility'.[24] He is keen to point out the specificity of this situation – enough travellers have stopped, and each is headed in the same direction – and also the very temporary constitution of the invention, which disperses as soon as the road is clear. He also suggests, and I will return to this in Chapter 7, that it is not just an invention that is produced when the boulder is pushed – it is at the same time a human rock-pushing collective that is in formation.

One of the more involved examples he gives is the invention of the Polaroid Land, the first camera to use self-developing film. If the initial mediation effected by earlier cameras was to enable light to 'work directly and automatically' on a photosensitive material or in other terms to establish a compatibility between the physical and the chemical, the Polaroid process inserted operations that had taken place in the professional dark room into the photographic apparatus.[25] Simondon recognizes that this is at once a technical invention, and also a social one, for it transformed picture taking from a professional activity to an amateur pastime. Moreover, taking a line of analysis that is unusual in his writing, he emphasizes that this invention was also driven by industry. Photography's decentralization from the professional darkroom was made possible by an increasingly concentrated relation with the industrial universe.[26]

Mediation – the possibility between hitherto incompatible orders of magnitude coming into communication – is at the heart of Simondon's complete system of individuation. In *L'individu* his examples are mostly limited to physical systems and energetic exchanges but in 'Imagination et Invention' the examples concern more everyday interactions with tools and technologies. Each time it is proposed that what is invented is not so much an object (a hoist, a camera) as a means of constructing compatibility that did not previously exist, that at once puts what were unrelated parts into a systemic relation with each other. We see yet more clearly, that Simondon's model does not assume a generalized connectivity already present between entities, but rather that it is through invention and individuation that relations between different orders of magnitude are rendered possible and come into being. In 'Imagination et Invention' Simondon extends his analysis beyond physical and energetic states to include social and even economic situations that might be brought into compatibility through invention. These examples suggest that Simondon's model of the 'complete system' is not just limited to form-taking but might be applied more widely. To return then to Barry's pharmaceutical molecules, we can say that what is invented here is not just a new chemical compound for a given service, but new mediating processes and mechanisms that enable these compounds to – for example – be shown to provide such a service (though certainly they will have other effects too) or to have the legal status of a patent. Information in this sense is not what enters the molecule, but what puts it into relation with other realities or 'orders of magnitude' that as Barry so presciently observes, have become central to pharmaceutical invention. This phenomenon cannot simply be defined with respect to the physical molecule itself, or even to the brightly coloured capsule that sits in its brown glass bottle. What differentiates one brand of pill from another may not be its physical structure, nor even its biochemical affects. It is the whole system making it possible to mobilize it as part of the pharmaceutical industry, from tests to legalization and the marketing of certain positive side effects over negative ones.

Simondon's 'complete system' thus provides us with a model in which all these steps that make mobilization possible are part of each system as it is defined, rather than prior to it, with mediation or 'information' at the centre. The benefits of a found herbal remedy gathered over time and spread by word of mouth 'invent' a relation between, say, the camomile flower and the promise of calm. This is a very different system, to that of the latest drug to come out of pharmaceutical R&D but nevertheless both remedies can sit side by side in my medicine chest. Similarly then, it is in terms of Simondon's 'complete system' that I propose we think of the very different material descriptions that we have identified in the architectural specification.

'That Constitutive Seam'

Choosing to look at the variety of forms of clause in the architectural specification has inevitably entailed grappling with the social, regulatory and economic context in which materials are mobilized for their use in building. These 'economic determinations', argues Frederic Jameson in his celebrated reading of Frank Gehry's own house in Santa Monica (1978), are precisely what must be concealed, even 'repressed' for architecture to be understood as art:

> Such materials clearly "connote"; they annul the projected syntheses of matter and form of the great modern buildings, and they also inscribe what are clearly economic or infrastructural themes in his work, reminding us of the cost of housing and building and, by extension, of the speculation in land values: *that constitutive seam* between the economic organisation of society and the aesthetic production of its (spatial) art, which architecture must live more dramatically than any of the other fine arts (save perhaps film), but whose scars it bears more visibly even than film itself, which must necessarily repress and conceal its economic determinations.[27]

The 'cheapskate' (then radical) materials Gehry used to build the extension around the original house – corrugated aluminium, steel mesh, raw plywood, cinder block, telephone poles – 'connote' as much the spatial invention, exposing the economic and infrastructural themes that architecture usually conceals.

If these 'scars' of architecture's production are usually concealed (and it is interesting that Jameson suggests that the appearance of hylomorphism contributes to their concealment) then Gehry's house makes them visible via the materials he uses, employing what we could call a 'critical materiality' to demonstrating their potency in the production of space. Importantly Jameson's words suggest that it is not in fact possible to sever aesthetic production from its economic context. In fact he proposes that speculation and the housing market are 'that constitutive seam' between them. Reading this in the light of Simondon's complete system, practices such as these can be seen in fact to provide the mechanisms for economic organization and aesthetic production to have some relation to each other. Reading this in the light of Sérgio Ferro's work, we might go yet further and propose that architecture's techniques are what render these two orders of magnitude compatible. At least for architecture in the present condition, these practices and mechanisms are constitutive.

Most discourse on materiality in architecture sits on one or other side of this seam. The so-called phenomenological architects and writers such as Juhani Pallasmaa, and Steven Holl, for example, turn to the sensual qualities of materials focusing on how they will be experienced and made sense of in the finished building, to sound, touch, smell.[28] Here the material becomes something beyond rationality but also beyond social forces. A 'material-based phenomenology', Nic Coetzer has written, 'is political in its *erasing* of any overt political traces'.[29] This is also the case with the 'truth to materials' tectonic approach of architects and critics such as Louis Kahn and Kenneth Frampton, who find in the material as physical stuff an innate logic that the designer is to follow. Even in Andrew Benjamin's explicit efforts to develop a materialist history of architecture that recognizes the historical nature of concepts, the potentials of his notions of 'material possibility' and the 'material idea' are limited largely to their effects on geometries. Materials are mobilized and enter into his discussion only in so far as they inflect form.[30] And despite its avowed attention to material change in architecture (and his wonderful introductory account of his formative experiences in the Hungarian Brick Brigade), Ákos Moravánsky's Semperian study *Metamorphism* has little to say about the effects of industrial and regulatory developments preferring to stay with the designer as the shaper of material.[31]

If for these commentators social, economic, regulatory and industrial factors are simply deemed irrelevant because it is the cultural, formal or aesthetic expression of materials they consider important, for others – and here I refer particularly to discourses of autonomous architecture – the industrial and technical aspects of architecture's materials are actively and intentionally treated as if they can be hived off. Famously Colin Rowe and Robert Slutzky replaced 'literal' transparency with a 'phenomenal' transparency that is not a property of materials but of formal arrangements (and, we might remember with a raised eyebrow or two, readable only to 'a man of moderate sophistication'[32]). Peter Eisenman dreams of a 'cardboard architecture' that can only be mimicked with a white no-hands construction when he realizes House I. In his effort to reduce the 'existing meanings' of the materials he uses, so that they are read not symbolically, but as formal markers in space he borrows the material palette of the International Style, in part because its planes of white or clear and grey glass are, he writes, 'closer to an abstract plane', even 'neutral'.[33] His marvellous essay 'Real and English: The Destruction of the Box' invents a mode of material analysis that achieves what seems almost impossible. Stirling's audacious use of brick and glass at the Leicester Engineering Building is interrogated entirely through structural logics and rules, with no mention of bricklaying or production, of the Englishness of brick or even the fact that it has been translated from handmade to precision material. Instead Eisenman mines the 'interior' of the architectural discipline in order to demonstrate the inventive and delightful ways Stirling plays critically with its material conventions. We can read 'Real and English' as an account of architecture's material play that at the same time (and brilliantly) voids it of its materials.

Too often then, architects and writers filter out the conditions of production and mobilization. In a rare instance of an architectural commentator noting the distinction between matter and materials, Robert McAnulty at once applauds the so-called material turn in architecture but at the same time pours scorn on most of its advocates whose practices he writes, 'were simplified to be those dealing with real materials, real constraints and real problems'.[34] Thankfully, however, spurred on often by the violence of materials extraction, to the damage being done to the planet and its inhabitants by materials production and by the mobilization of building materials to dislocate settlements of poor and indigenous peoples or dismantle the power and autonomy of building workers, more researchers are studying their production, consumption and life cycles. Mark Jarzombek's 'The Quadrivium Industrial Complex' makes the case for this research in polemic terms with respect to the environmental devastation caused by our attachment to the four primary materials of modern building – steel, concrete, glass and plastic.[35] Architectural studies of concrete are least likely to fall into an entirely aestheticized discourse with Reyner Banham famously arguing for the primary role of fire and US insurance companies in the widespread take up of concrete in the early 20th century.[36] Making a very different case about turn of the century France, and drawing on Adrian Forty's discussion of Hennebique's office in *Concrete and Culture*, the Brazilian theorist and historian of architecture and labour, Sérgio Ferro, argues that the take up of concrete was in fact a means to displace skill from the construction site to the drawing office of engineers, thus breaking the considerable power that construction workers and unions had held. In Ferro's account, modernist architects and critics were merely capital's ideologues with their avant garde manifestos and novel concrete forms.[37] Importantly, Ferro also credits the increasing detail and precision of the specification as a tool in that displacement. Geographer Matthew Gandy's seminal study of the reworking of raw materials to build New York,[38] and landscape architect Jane Hutton's brilliant tracing of the material translocations and their impacts in constructing five more recent sites in the same city[39] are models for critical research into the chains of material processes through which the built environment comes into being. Some of the most interesting studies looking at regulation and specification in relation to building materials explore how they have been used as racialized tools in colonial contexts. For example, Robby Fivez and Simon De Nys-Ketels look at how a law on building materials in the Belgian Congo was used to enact racialized urban segregation, Nic Coetzer has made similar studies in South Africa, and Lisandra Franco de Mendonça in Lourenco Marques, Mozambique.[40] In 'Bad Earth' Hannah le Roux and Gabrielle Hecht show how the manufacture and testing of new cement blocks in postwar Africa, often under the guise of self help and development, 'circumvented the need for skilled local builders and their ability to create durable structures' themselves, channelling money to building materials companies and products like asbestos.[41] By taking a much wider and more holistic view of the lives of building materials and how they are put to work these studies, among many others, demonstrate the multiple relations that can be engendered via materials.

In choosing to look at materials in architecture through the specification, I have of course made it impossible to ignore this 'constitutive seam' of architectural production. It might be argued then, that my choice of method already predetermines my conclusions and, to some extent, this is the case. The specification is necessarily embedded in the statutory, contractual, and industrial worlds of making buildings and its descriptions of building materials will reflect that. But I have also tried to show that how a material is prepared or made ready for building has some effect on what it is capable of doing or – more precisely – on what is able to be compatible with. To think of building materials in terms of Simondon's complete system, is to recognize first that the preliminary operations and the preparation of potential are part of the system, not to be hived off as prior to or outside the material's form-taking. Moreover, it is to put first the question 'what is the material capable of becoming compatible with?' ahead of the question of what it is as a substance, or even what it does.

In the case of some building materials, as in the guide specification for putty below, it may simply be that the material is defined by its capacity to be compatible with another material:

> Putty is to be linseed oil to BS544: 1969 *for glazing to wood* and approved best quality metal-glazing compound *for glazing to metal*.[42]

In other cases, such as this recommended specification for hardcore (1985), we have seen how materials are set into relation with systems of measurement:

> Make up to required levels under concrete beds and pavings with approved brick hardcore *broken to pass a 75mm gauge*.[43]

In the case of rigid formwork, the chains of macro- and micro-processes involved in concrete production establish the possibility of communication between the liquid concrete and its timber mould, but following Simondon, we could go further and suggest that what is constructed here is the compatibility between two more disparate 'orders of magnitude' the sluggish mass of concrete and the designer's geometrically conceived and drawn idea.

It is with performance specification and engineering and the corresponding regimes of testing, quantification and homogeneous material production, that building materials rather suddenly become capable of compatibility with a wider range of realities beyond the physical processes of building and design, from thermal comfort to security threat to the cultural imperative to be clean. We should not be so quick to assume, as Benjamin or McAnulty do, that the realm of 'architectural effect' is (at least today) any more immune from 'real materials, real constraints and real problems'. By ignoring them, don't we remain blind to their effects, and unable to intervene in them? Is it acceptable that the materials we work with are enlisted to do things we might not ourselves want to take part in? Or that through them we conform to sets of social parameters that we have had no say in determining? Moreover, we might want to ask what possibilities might emerge out of these developments beyond those established within normative practice. How might we play with thermal comfort or extend the relations between building materials and subjects beyond those already defined by industry?

We can identify very different regimes of specification practice at work in today's building industry. On the one hand there are the performance-oriented specifications produced by companies such as those I have represented with Schumann Smith. But on the other we also see specifications for materials and components of building that could not possibly conform to the descriptive constraints of open specification. Some even confound the mix and range of forms of clause that remains available in the NBS. In both cases it seems to prove impossible to conform to any one mode of specification.

In their open specifications Schumann Smith have devised descriptions of materials that are intended precisely to avoid the prescription or naming of specific materials, and to leave selection of particular techniques and products to the contractor. They call this (non-prescriptive) form of clause 'descriptive'. In the majority of cases descriptive specification is achieved by stating performance criteria, but it is not always possible to keep to this form (nor is it always desirable). From time to time even full-blown process-based clauses creep in. Here for example are some very detailed clauses from a Schumann Smith specification giving the processes to be carried out in the removal of bird droppings from stone-clad façades:

C40.3102   Bird Guano Removal

a)   Prior to cleaning any stone, all areas of bird dropping build-up (not to be confused with bird dropping staining which refers to light coloured staining on vertical surfaces) shall be removed with wooden or plastic scrapers. These areas shall then be cleaned with hot water and stiff bristle brushes . . .

C40.3104   Bird Staining Cleaning
            Before cleaning the facades, all bird dropping staining on all facades shall be treated in the following manner:

            a)   Wash the surface thoroughly with high-pressure hot water
            b)   Thoroughly wash the entire area with pressurised water (450 psi)
            c)   Repeat the process as required to fully remove all stains
            d)   Consult with the Employer's Retained Architect in connection with any stains not removed by the specified treatment[44]

Despite the efficiencies of façade engineering and progressive specification the business of cleaning up after the birds still requires scrapers, stiff bristle brushes, hot water and elbow grease. It is still a straightforward labour, a process that acts on the material and is unavailable to quantification. In the section for glass façades, the cleaning requirement is included in the clause specifying 'Durability'.

> H11.1424 Impact and Abrasion Resistance
>
> a) The works shall resist abrasion from agreed cleaning methods and maintenance systems without any noticeable change in surface appearance.
> Generally, surfaces shall be sufficiently hard (including glass coatings) to resist all reasonable impacts from hand-held objects in accordance with BS EN 358.[45]

All that is said is that the material is to be sufficiently hard so that it 'resists abrasion from agreed cleaning methods . . . without any noticeable change in surface appearance'. Cleaning is anticipated in the clause, but only in as much as it is not to affect the material, in particular, unsurprisingly in relation to the way it looks. A self-cleaning glass might be specified here, but at least at the time of writing this, no quantitative methods have yet been devised and standardized that would enable that glass to be performance-specified.

As we saw in Chapter 5 with respect to the specification of natural stone, some materials resist performance specification because they occur naturally and cannot be assumed to perform in the same way from sample to sample. Schumann Smith's clause for quarried Portland Stone is to all intents and purposes prescriptive. The stone is named by quarry as is usual, but it is given as an 'indicative material'. If the 'indicative stone' does not in fact conform to the other criteria given in the specification it is the contractor's responsibility to find another stone.

> H51.2000 Systems/Materials and Fabrication
> Eg. H51.2101 Portland Stone Type STN-02
>
> a) Oolitic limestone to BS5628 Part 3
> b) *Indicative stone:* Coombefield Whitbed
> c) Quarry: Coombefield, Dorset
> d) Colour: Buff white, to match and fit with accepted range samples
> e) Texture: fine grained
> f) Stone sizes. Large panel sizes up to 1500mm x 750mm as indicated on the Design Drawings *or as otherwise determined to meet the appearance and performance criteria.*
> g) No saw mark shall be visible on the finished surface
> h) Stone selection and testing to comply with the CWT 'Guide to the Selection and Testing of Stone Panels for External Use'[46]

In this 'indicative' clause Coombefield Whitbed has a curious status. It is named, its geographic source is given, but it is put forward as a kind of idealized material to which the designer aspires, which can be used by the contractor if it satisfies the given criteria or substituted by another to satisfy the same criteria including appearance. A caveat in the section defining descriptive specification explains that in fact in all cases, 'Where a particular material, product or supplier is indicated in the Specification such material, product or supplier shall be deemed indicative representing the Employer's Retained Architect's design intent only'.[47] Here, another curious category 'design intent' has crept in. In a peculiar way, and arising from the contingencies of claims-proof contractual processes 'design intent' is something like the critical architect's aspiration for a purely formal architecture. It is design that meets performative and visual criteria given by the architect, but with the specifics of its material realization emptied out.

When architects hand over the detailed design of a part of the building to a specialist subcontractor but still require some control over its visual appearance Schumann Smith augment performance clauses with 'visual' or 'design intent', a new form of clause which ties the design drawings into the specification. Taylor explains that design intent is only relevant for the visible parts of the building. Hidden structural steelwork would require only engineering input, whereas exposed steelwork would have 'architectural requirements' and would demand the specification of visual intent. This is a peculiar division. It leaves the architect's role as designer of the visible. The architect designs the form and visual appearance of the building and gives the performance criteria, whilst the fabricator or engineer ensures that it conforms to these criteria within the parameters of visual intent. We might for example see the wood panelling which Diller & Scofidio + Renfro designed for the Alice Tully Hall at the Lincoln Center in New York as a case where the material is specified in relation to visual criteria but engineered to meet performance criteria.[48] The panelling had a complex undulating surface that needed to meet a range of visual and behavioural criteria along its length, described here by Ruben Suare, the lead architect at 3form who manufactured the composite wood and resin panels:

> One-inch thickness, a class-A fire rating, approval by MEA (Materials and Equipment Acceptance) to address toxicity requirements, formable to various complex geometries informed by acoustic requirements, no difference between solid and translucent panels when not lit, and capable of sequencing the wood throughout the concert hall.[49]

For the architects, wood was to look like book-matched timber panelling whichever way it functioned:

There are intimacy issues, trying to get everything into the hall, and doing it all with one very strong and versatile element, and that is wood. Wood can be steps, wood can do all the sound shaping, and wood can produce the effect of the enveloping quality of light.[50]

In terms of the material intentions for this choice of material, it is the look of the material that marks it out as 'wood' for the architects and must be consistent, while its properties can vary and are given in terms of performance criteria. In a sense, then, the traditional form/matter model is replaced here by visual intent/performance, in the same division that structures the Schumann Smith specifications. Performance requirements are usually determined by building regulations and industry norms. The architect need only know how their design will be used and inhabited to select the appropriate criteria. Their job is simply to ensure their design is compliant with performance requirements that are normative and given elsewhere. The requirements are parameters for the design, more than part of what is being designed. In the Alice Tully Hall architectural intent is little more than 'the wood look' set up in such a way that performance requirements produce its variegation rather than any prior processes of organic growth. To recall Barry, it is not so much that information 'enters' this material. Rather this material grows from information. Its composite manufacture allows performance criteria to determine its thickness, its curvature, its translucency and its effects.

Open specification restricts the architect's access to decisions about the materials to be used in building and decouples knowledge about materials, their provenance, their lifecycles, how they are worked on site from the process of specifying and designing. As well as promoting the use of certain kinds of materials produced at large scale that are able to demonstrably meet the criteria, open specification prescribes a particular kind of design in which materials are 'indicative' and approximate an abstracted, idealized matter. The architect might want to consider issues of social and environmental justice and take what Pauline Lefebvre has called 'responseablity' for their material selections, but they can operate only within the parameters of 'design intent' and formal manipulation. The contractual context of a neoliberal building industry, rather than their theoretical position and ambition, has rendered them purveyors of Eisenman's 'cardboard architecture'.

The open specifications of Schumann Smith represent one end of the contemporary spectrum. They work only for materials that have been made ready for this system. At the other end are materials and components of building that could not possibly be described within the constraints of such documents, nor even within more traditional forms of specification. To include these materials architects may have to devise techniques that go beyond the use of conventional documents. First, where materials cannot be prescribed through generic clauses, specifications may still be carefully written out in full by architects. For example, if unconventional construction is used it may mean that the specification is written from scratch without reference to a master specification as when Caroe and Partners Architects designed a 'primitive hut' to be built in a garden which used saplings stripped of their bark but otherwise untreated. The specification René Tobe produced on their behalf is refreshingly prescriptive and invents its own rather detailed descriptions:

> 4.3.2   Structure to be exposed columns and bracing members of solid timber saplings with bark stripped as shown on drawings. Vertical columns to be 8 no. 160mm diameter oak saplings with bark stripped columns. Diagonal and horizontal cross brace timbers to be 100mm diameter oak saplings with bark stripped brace timbers.[51]

Second, it may be the case that most parts of the building can be described within the parameters of a master specification such as the NBS, while some construction methods exceed conventional practice. The NBS may not have added the most recent applications of materials, and was unlikely to include innovative one-off techniques. For example, in the specification for the Potters' Fields Park pavilion (2006) DSDHA described their innovative charring technique by adding their own clause into the main body of the NBS framework as follows (note that this is given as an 'indicative method'):

> **H21   Timber Weatherboarding**
> 01   Description:
> –   Work in this section, together with glazing, rigid sheet cladding, doors, etc. will be subject to Contractor design.
> –   Bespoke timber weatherboarding system with 'blackened charred' finish to be developed with Architect (see appendix A – indicative method)[52]

They included further details in Appendix A, where a series of process-based clauses describe the production of samples for approval by the architects in unusual detail:

> 01 **Description of Tests:**
> – Provide *two* charred sample strips, prepared in a controlled environment of each ~~of the three timber types~~ in 1.2m lengths with the cross section as shown in diagram 1 below. These timber strips will be charred using ~~two different~~ the following method.
> – Method 1
> Method 1 will use a high performance hand held propane gas burner with an output of 125 KW, see image 1 below. Each strip will be hand charred using this burner and will then be ~~sandblasted or dryiced~~ waterblasted to remove loose surface char. This will bring the surface to a 'clean to the touch' finish with the appearance of fire damaged timber. One sample is to be subjected to 'intense' waterblasting and the other to 'light' waterblasting in order to ascertain the appropriate aesthetic.[53]

Rather delightfully, a photograph of the recommended propane gas burner (Fig. 6.1) is included with this appendix.[54]

Here the architects avoid outright prescription by referring to their innovative method as 'indicative' within the main body of the edited NBS and by giving the process-based details of the method outside it in the kind of attentive language we might recall from Val Harding's specification for casting

FIGURE 6.1 DSDHA Potter's Fields Park Pavillion, Image of high performance burner appended to specification. Source: Courtesy of DSDHA.

concrete walls. It is interesting to note also, that they refer to the achievement of an 'appropriate aesthetic' – which suggests the centrality of 'visual intent' without utilizing the formal language and specification structure used by Schumann Smith. DSDHA are able to adapt the NBS to specify their innovative material techniques but Cottrell & Vermeulen Architecture, a practice who integrate the use of salvaged materials into some of their building projects, prefer not to write specifications at all. Instead, they work closely and usually with the same contractor, ensuring quality and workmanship through an ongoing relationship of trust, which makes the need for formal specifications redundant.[55] In this case material innovation is possible because of contractual arrangement that itself lies outside normative practice. The Stock Orchard Street 'straw bale house' designed by Sarah Wigglesworth Architects (SWA), used a range of materials and techniques, many of which were invented specially for the project such as cement-filled sandbags, an upholstered wall, thick gabion 'columns' and, of course, straw bales. Many of these were outside those covered by the standard clauses of the NBS and what building regulations defined as 'proper' materials (Fig. 6.2):[56]

29.

Building Regulations – Regulation 7

Any building work shall be carried out:

(a) with proper materials which are appropriate for the circumstances in which they are used; and

(b) in a workmanlike manner.

"Proper materials" shall include materials which:

(a) bear an appropriate EC mark in accordance with the Construction Products Directive; or

(b) conform to an appropriate harmonized standard or European technical approval; or

(c) conform to an appropriate British Standard or British Board of Agrement certificate; or

(d) conform to some other national technical specification of any member state which provides, in use, an equivalent level of protection and performance with respect to the relevant requirements of the Building Regulations, as an appropriate British Standard or British Board of Agrement certificate.

A:\9-stock-orchard-street N7.doc/bc-regs/sh/july98

FIGURE 6.2 Sarah Wigglesworth Architects, Stock Orchard Street house. Schedule attached to letter from London Borough of Islington. Source: ©Sarah Wigglesworth and Jeremy Till.

Notably, proper materials are determined by their conformity to national or European standards and should have been tested to demonstrate this. SWA produced very thorough detailed drawings but no specification was written as part of the construction documentation. In part this was because many of the techniques had to be worked out on site rather than prior to construction, which in turn demanded a different relationship with the contractor and builders. Details and costs could not be finalized before building and the builder was often involved in developing construction solutions. In the case, for example, of the building of the sandbag wall which faces the railway track running along one side of the building, the architects designed a complex rig which could be used to fill the bags with cement (Fig. 6.3).

In the event, the builders designed their own ad hoc system using items already on site such as plastic piping. According to Wigglesworth and Till this kind of inventive involvement in the building process gave builders a different relationship to it:

> In their making of the building, the builders have suspended their initial disbelief in the project and have claimed the various unknown technologies as their own.[57]

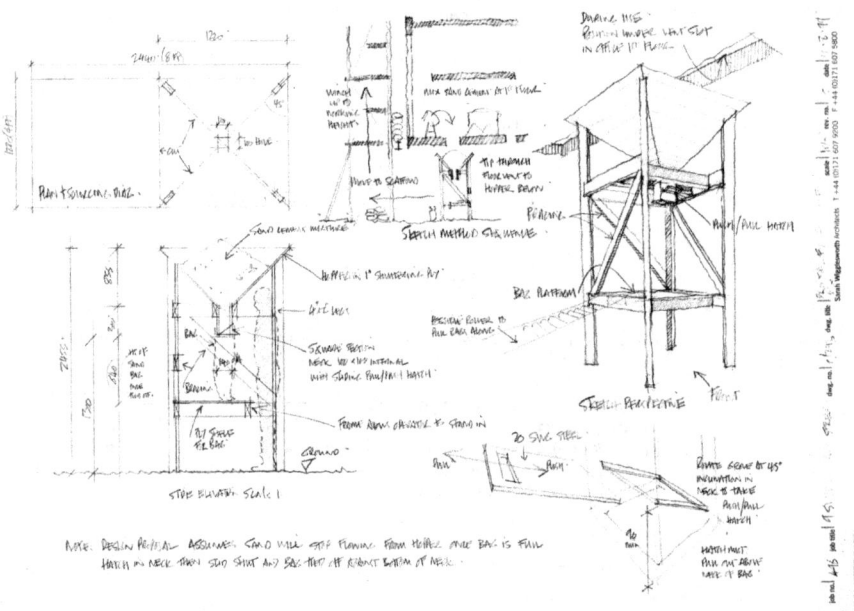

FIGURE 6.3 Sarah Wigglesworth Architects, Stock Orchard Street office. Construction drawing of sandbag-filling rig. Source: ©Sarah Wigglesworth and Jeremy Till.

More significantly, perhaps, the fact that the sandbags could not be included as a 'proper' material in the terms of the building regulations meant that their performance was not officially considered as part of the functioning build-up of the wall (Fig. 6.4). Although the bags dampened sound and vibration from passing trains an inner wall built with conventional materials had to be demonstrated to perform this role. At least in terms of compliance with building regulations, the sandbags became 'extrafunctional'[58] and their status was relegated to that of mere decorative cladding.[59] The physical properties of the materials are thus deemed irrelevant, in as much as they become extraneous because of the mechanisms of statutory approval.

Similarly, in the case of the well-known straw bale wall, only some of its functioning capacities were recognized as contributing to the approved performance of the wall. The Building Control officer decided there was not sufficient evidence to allow the bales to be used structurally, and to avoid their outright rejection the architects designed a vertical timber truss which took the official, demonstrable role of taking the loads (Fig. 6.5):

> The argument is that the wind loads are transferred via the cladding ladders and therefore the bales do not enter into the structural argument because they are being treated as non-structural infill material.[60]

In this case then the bales become 'extra-structural' because of their statutory status rather than the physical capacities they in fact possess. But when it came to providing an agreement certificate to guarantee the viability of the straw bales as a building material, Building Control accepted a batch of alternative documentation compiled by the architects from straw bale builders' handbooks, websites and informal advice. In lieu of the usual statutory and industry authorized tests and standards, the voices of amateur fabricators entered the negotiations around the 'proper' material and all parties involved, including Building Control, drew on their experience and capacity for independent evaluation, rather than the simple regulatory mechanisms of inclusion and exclusion. At the Stock Orchard Street house, the degree to which the performativity of the sandbag wall or the straw bales is recognized is a result of the statutory mechanisms which can include or exclude performance as a relevant attribute of a given material, just as in the specifications produced by Schumann Smith and Allies and Morrison where the use of performance specification will depend on factors such as the architects' expertise with the material, its location in the building, the contractual arrangement it is part of, as well as the kind of material it is (natural, standardized, mass-produced, performance-engineered and so on).

Just as performance specification is part of a specific configuration of practices, techniques and material technologies, and other forms of clause

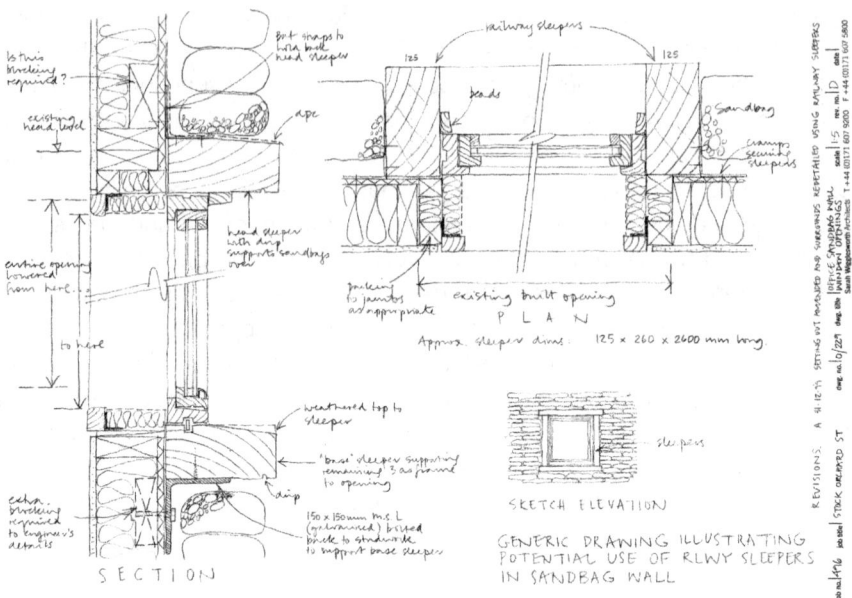

FIGURE 6.4 Stock Orchard Street office. Detailed section through sandbag wall showing inner performing wall. Source: ©Sarah Wigglesworth and Jeremy Till.

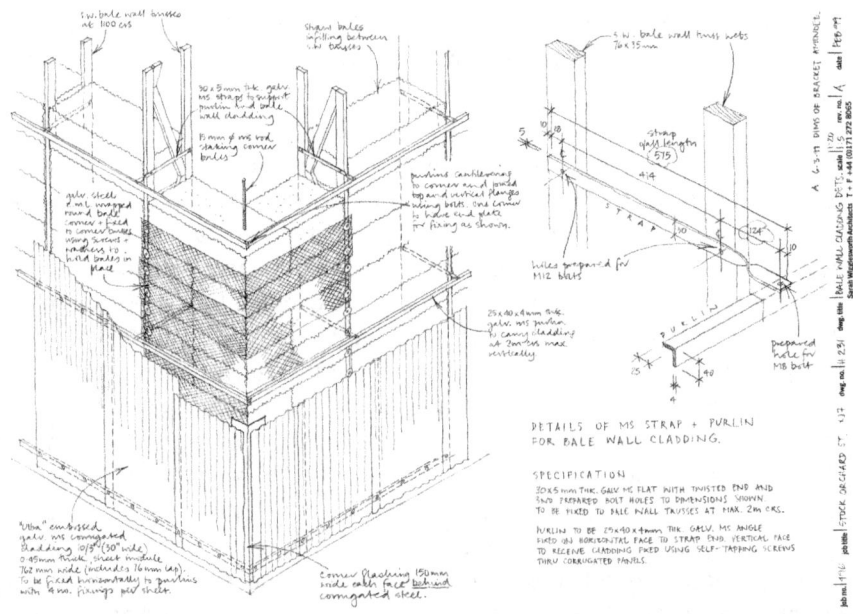

FIGURE 6.5 Stock Orchard Street house. Axonometric section through straw bale wall showing structural 'ladders'. Source: ©Sarah Wigglesworth and Jeremy Till.

are evidence of other systems, we can also identify systems of material that cannot be included within the master specifications currently available. If we look at these we see that different practices and techniques prepare these materials for their mobilization in building, ranging from ad hoc or inventive specification practices to abandoning the use of specifications altogether. A cement-filled sandbag as used at the straw bale house, which has not been part of this regime of evaluation might have the physical capacity to insulate against sound and vibration but these capacities cannot be recognized with the regulatory framework. The form of clause used in a specification (or indeed the absence of a written specification) has a bearing on how any given material can be mobilized. It is one of the factors (or 'preliminary operations') that prepares a system of material, and differentiates it from another, even where the same physical substance is being prescribed.

To see materials in terms of systems is to include the processes that prepare them for use as part of what characterizes them and makes possible our uses of them. This has a number of benefits. A process-oriented understanding of materials can be helpful in an era of global warming that demands we make more careful use of our limited energy supplies and resources. Now that performance standards are often written into the materials and design technologies we use, and materials production is enfolding an ever-widening range of criteria into its processes a model of materials that encompasses these factors can give architects a greater awareness or control of them than one which endeavours to hive them off. To see materials as systems gives us more nuanced, or 'concrete' ways of understanding the differences between them. It enables us to select materials not just for their finished appearance or their properties, but also for the kinds of regulatory frameworks and procedures they will or will not set in motion, and the kinds of relationships that might be engendered on and off site. These issues would not be characteristics of materials taken as simple physical substances, but they would be included in a conceptualization of them as 'complete systems', that included the preparations for their mobilization and the kinds of relations they engender.

Moreover, to consider building materials as systems in Simondon's sense more than as substances also provides an explanation for the variety of forms of clause. With the exception perhaps of the earlier dimensions-based specifications that were in use before the drawing began to play such a regular part in building documentation, it is rare to find a specification that makes use of only one form of clause. Despite the intentions of the team who developed the NBS to move towards the description of 'ends rather than means' this master specification still uses an assortment of forms of clause, and even Schumann Smith fail to describe buildings entirely in terms of open 'descriptive' specification. In fact, the imperative to do this just results in the invention of yet more forms of clause. If forms of clause were simply alternative modes of description, one available in an era of artisanal production of buildings, with another taking over in an era of design and build, we would expect to see any deviation from the prevailing form as a peculiar exception. Instead, variety is to be expected – there is a *persistence* of variety in the forms of clause. And although new forms of clause tend to emerge – at least with any force – when industrial and contractual conditions change, they also tend to 'stick' to specific material production processes. It is difficult to specify mortar and in-situ concrete through any other means than the process-based clause. Naturally occurring materials will tend to be named – even if this naming is almost avoided by using the term 'indicative'. There is something decisive both in the physical nature of the material and the demands of its working, and in the material and immaterial processes that prepare and define it for its mobilization in the building industry. If we propose that these constellations understood in their complete sense are decisive more than some essential physical arrangement of the material, or its discursive constitution in one epistemological framework or another then Simondon's complete system becomes a productive way to formulate these material constellations.

The different forms of clause in the specification are evidence at once of the very different ways a material may be mobilized in building, and of the different regimes of preparation that make that mobilization possible. Where the process-based clause is used, the system of the material includes only its fabrication. The preparations described are mostly physical, to be carried out on site or, on occasion, in the factory. Where the performance clause is used the system of the material extends to the way it is intended to behave in the finished building. A much wider range of preliminary operations is required to make it possible to specify a material in this way. The form of clause is an indicator of a specific system of material. It is also part of the preparations that make that material ready for its specific mobilization in building. Its use is dependent on those preparations. In recognizing the variety of forms of clause used in specification practice, and the means by which materials are mobilized in construction where the document is not used, the plurality of systems of materials in building becomes apparent. Not only do the systems employ different techniques and materials; there is for each the 'constitutive seam' within which the system emerges, and in turn makes possible.

7

Going into the Mould

Of the three forms of clause explored more closely in this book, the process-based clause has had a certain priority. Its detailed descriptions of the preparations for mixing mortars in 18th and 19th century specifications, and for the mixing and finishing of concrete and the making of formwork in 20th century specifications appeared analogous to Simondon's drawn out account of the form-taking of the wet clay brick. Once I started to think his ontogenetic philosophy might also yield insights into the thinking through of the other forms of clause, despite that both naming and performance specification exclude process, I ordered Simondon's *L'individu*. But when my copy arrived – a first edition – the book could not be opened. The tops of the pages and many of the long ends were still joined together. The book had not yet been trimmed. Like the example Simondon gives of 'a book used as a wedge or a pedestal' that 'cannot bring information',[1] this copy of *L'individu* had the potential to be read, but it was not yet ready to come into relation with its reader.

The book had been produced in quarto format. Sixteen pages had been printed on both sides of a single sheet that had then been folded three times and bound. Even to begin reading *L'individu* was a slow process that involved making three cuts (two short and one long) with a paper knife every eight pages – one hundred and forty-four cuts in total. This was how books used to arrive with their readers and they too cut them open with a paper knife. The margins of my copy are wider than they would have been had they been mechanically trimmed, with more space for my pencil notes, and because my cuts were rougher than if cut with the blade of a trimmer; the book's page ends are velvet soft to the touch.

In the light of the arguments cocooned in my untrimmed copy of Simondon's *L'individu*, the cuts I made one by one can be understood as the last sequence in the chain of operations that make a book ready for reading (from the felling of trees and the pulping of paper, to the development of written language and printing techniques, and in my case, the French dictionary always at my side). These preliminary operations are usually out of sight, but, on this occasion, I participated in them and became aware of them, before the dynamic encounter began with Simondon's thought. I had been obliged to (in Simondon's terms) 'create the conditions' for my own reading, and in doing so I had for a moment entered 'into the mould' where the processes of individuation become visible. It is only in the closing chapter of *Du Mode* that Simondon finally explains why we so urgently need to develop an understanding of technical objects that goes beyond their usefulness to us.[2] We need, he argues, to find other ways to live and engage with technical objects and ensembles than estrangement and alienation. In the 20th century 'the human world of technical action has once again become a stranger to the individual'.[3] When we labour with tools, we see only the end goal. Our work creates the conditions for the technical operation to unfold, but at the same time obscures it. We neither know it, nor have the capacity to intervene with it. Working with machines reveals this alienation yet more clearly. Simondon's insights into our lack of technical culture resonate with architecture's dismissal of technical knowledge and practices. We need, he insists to develop technical knowledge, which 'consists in starting from what happens inside the mold in order to find the different elaborations that can prepare it by starting from the center'.[4] 'Going into the mould' is for Simondon a means both to know the technical operation and to come into a communicative relation with it – which he terms invention.

Preliminary Operations

To a great extent, the conclusion of this book has already been reached. Our starting point was to consider materials not so much as substance or force, nor even in terms of Aristotle's 'ousia sunolos'– the combination of matter and form – but rather in terms of their ontogenesis or coming into being. In his book *Concrete and Culture* Adrian Forty makes this point with respect to concrete, which can, he writes, 'be more accurately described as a *process* than as a *material*'.[5] Indeed, he goes on to suggest that this insight might equally be applied to other materials whether stone that must be quarried and hewn, or timber that must be transformed through human labour for its use in building. All materials must be worked or prepared for building. Through close attention to the variety of forms of clause in the architectural specification, this book has extended these material preparations to include a full range of operations, from physical processes to classification to the quantification of material properties and through testing their rendering equivalent to environmental conditions and the comfort and actions of human users.

In addition, each form of clause pertains to a different system of material, which in turn contributes to the constitution of a material's possible mobilization. As we saw, for example, at the Stock Orchard Street house (in

Chapter 6), the physical properties of the cement sand bag wall and the padded upholstery insulate the office building from the noise and vibration of passing trains, but because they have not been performance-tested and quantified they cannot be performance-specified. As such their insulating properties cannot be mobilized in so far as building regulations are concerned, and a second wall using conventional construction that can demonstrably be shown to perform in this way must be built. Within this logic the outer wall becomes 'extra-constructional', supplementary and 'decorative' even despite its actual performance. Here we extend Simondon's notion of individuation to include not just how a material comes into being, but also to the possibilities of its use in building. In process-based specification this almost always finishes at the point of construction, as indeed in Simondon's account of the clay brick individuation ends at the point of form-taking, whereas in performance specification it projects forward to the material as it will behave once built. Within the classificatory practices of naming there may be traces of its ontogenesis (as in the practice of naming timbers by where they were traded) or indeed of its potential properties (hardwood, softwood) but naming posits the material as a substance that can simply be transposed from one situation to another without inflection.

In this sense, then, the specification of a material – or more precisely, the form of clause through which it is specified – is also part of the preliminary operations for any given system of material. The choice of clause can grant a material a part to play in securing against a terrorist threat or designate it as a branded commodity within a system of exchange. And in a few rare cases, the specification acts as the only preparatory factor in the material's mobilization. Here, tucked away in the paint specifications for the Elfrida Rathbone School (1961) for example, is a clause specifying the paintwork for a steel column:

26.13 ...Each succeeding coat of priming and undercoating paint shall be sufficiently different in colour to be readily discernible ......[6]

At first we might wonder why Clause 26.13 specifies each layer of paint to be a different colour, even though the colours will be hidden (and only seen, perhaps, when one day a child rides her bicycle into the column and reveals the rainbow of colours beneath the surface). The specification of paintwork is usually concerned with appearance and maintenance. At the physical scale each layer applied to the last enables a transition from a specific material (wood, metal or plaster) to a more general finish (gloss, matt and so on).

Texture is given in rare cases, such as the clauses below specifying intumescent paint finishes:

M61	INTUMESCENT COATINGS FOR FIRE PROTECTION OF STEELWORK
440	BASIC FINISH
–	Definition: Reasonably smooth and even. Orange peel, other texture, minor runs and similar minor defects are acceptable.
450	NORMAL DECORATIVE FINISH
–	Definition: Good standard of cosmetic finish generally, when viewed from a distance of 5m or more. Minor orange peel or other texture is acceptable.
460	HIGH DECORATIVE FINISH
–	Definition: High standard of evenness, smoothness and gloss when viewed from a distance of 2m or more.
490	TOP SEALER COAT
–	Application: To achieve dft recommended by manufacturer and to give an even, solid, opaque appearance, free from runs, sags and other visual defects.[7]

Even the application of layers of paint could be understood in terms of form at the micro level. Indeed the architect Adam Caruso does this when he describes the 'microtopography' of the overpainting over decades of the aluminium panels at the Van Nelle factory (Rotterdam, 1931).[8] But in the Elfrida Rathbone specification the rainbow of colours is mobilized as a checking device. The different colours are there, presumably, to allow the architect to check that each layer – primer, undercoat, topcoat – has been applied as instructed. The paint is prepared for the architects' visit, for an 'extra-physical' demand that is materialized in the fabrication and becomes a physical part of the column. What alters here is not so much the physical make-up of the material, as the 'system of material' that makes possible the relationship between the paint and quality control. This work could also have been carried out by other means (by a social system of trust for example, or through surveillance CCTV). Here it is achieved through the mobilization of paint in this particular system of material.

Clause 26.13 exemplifies beautifully that it is not just the physical make-up of a material or its aesthetic properties that constitutes a material's capacity in building. It is by catching sight of *this* build-up of paint, in the moment of being mobilized for its potential to become part of the architect's procedures of approval that we can see them emerge together as a system of material. Moreover, in this unusual case the specification itself is the sole preparation that creates the conditions for the various shades of paint to be mobilized in this way. This one instruction sets up the painted column in such a way that there can be compatibility between the material and the architect's routine checks on site. This potential is only actualized in the fleeting moment of checking. Once the architect's check is complete the rainbow residue is still there, but redundant.

The chains of operations that prepare any system of material for building are typically far more complex and prolonged than this; each system has its own specificity. The preliminary operations set out in the clauses for in-situ casting at Val Harding's house prepared both concrete and formwork for a form-taking that was to follow the architect's smooth orthogonal lines. At the Elfrida Rathbone School they prepared a system in which form-taking followed instead the micro variegations of the sawn timbers' grain. The creation of conditions in which performance-engineered glass can be mobilized involves complex procedures of quantification and often-cumbersome tests. These preliminary operations enable films and depositions at the micro-scale to be mobilized for their demonstrable capacities to reduce heat transmission (and enhance the comfort of an executive in a light shirt) or resist the impact of specific weapons (dependent on local and global levels of threat).

The selection of one form of clause over another has effects on how materials are conceptualized and mobilized in building. Paying attention to these preparatory effects suggests that a material's genesis and the plans for its use are determining. We would better approach materials as 'funeous' systems rather than as substances. Just as Simondon identifies various modes of individuation from the technical to the physical to the living and the psychic, the forms of specification – naming, process and performance specification – indicate various categories of material system. Following Simondon's conception of the system of individuation means understanding what is made ready through the preliminary operations are the conditions for relation. It is their very different capacities for relation – whether between a malleable material and the idea of form or between a sheet of glass and a comfortable efficient workforce of between a strata of colours and contractual approval – that are most distinctive about each of these systems of material.

For this enquiry into building materials, following Simondon's ontogenetic approach has enabled us to see that the relations constituting individuating systems are themselves constructed. They are, in Simondon's words, 'veritable relations' rather than 'a simple connection between two terms that could be adequately expressed using concepts'.[9] Moreover the relation is to be 'understood in its role as relation in the context of the being itself, a relation belonging to the being, that is as a way of being'.[10] It is not to be understood as pre-existing the being, nor can it be abstracted. Simondon asks us to look at the specific coming into being of the compatibility that an individuating system is organized around, that is at the same time its individuation. This applies to systems in each of the regimes he studies, from the physical; the crystal seed that discovers compatibility already latent in the preindividual state of a crystal solution, to the vital, and to the psychosocial; where for example it is only through the common effort of a group of travellers that each individual's bodily force is capable of having effect on the boulder. Their individual relation with the rock comes into being at the same time as the group temporarily becomes a rock-pushing collective.

With this in mind, an ontogenetic theory of building materials would study systems of materials over materials as substances, and attend in particular to the distinctive relations made possible by each system. An ontogenetic theory of building materials would do more than replace an account of passive matter, with one of active, 'vibrant' matter or materials-as-process. It would ask how contemporary developments alter materials' capacities for relations beyond their physical properties, and seek to understand the chains of preliminary operations that create the conditions for the possibility of new compatibilities. An ontogenetic approach to building materials requires architects and critics to engage beyond design and architectural discourse, with the contexts in which materials are prepared for mobilization. This would include (with respect to the concrete / mould relation for example) following and becoming aware of processes from extraction to the building site, and also the wider factors though which shape materials in so far as they are understood as systems – such as the building industry, manufacture, law and regulation.

For example, the fact that performance specification demands regimes of testing and homogeneous production that can be carried out by large-scale industry, means that techniques of performance specification also exclude all kinds of building materials and material practices that cannot be incorporated into that regime. The apparently inconsequential imperative to describe materials in terms of performance, in fact influences which materials are at our disposal, and steers us towards those produced by the global industrial complex. Moreover, the performance apparatus also facilitates the invention of an ever-increasing range of compatibilities between materials and the users of buildings and their behaviours, and the social and political milieu. These new systems of material have architectural effects, which we should be aware of. However, they are produced and enshrined in standards that are outside the architect's remit. In Simondon's terms, the chains of operation that engineer these compatibilities are 'preliminary' and as such they are out of sight. They are left in the hands of regulation and manufacture and well beyond architects' control. By ignoring developments such as these because they are 'merely technical' – or outside architecture 'proper' – we remain blind to some of the most insidious ways that materials are being put to work. Moreover, we abandon any potential they could offer, handing them over to normative industrial practice without asking what else we might do with them or what else they might make possible. For Simondon, it is not enough to know and intervene in the preparations for technical relations. The possibility of ongoing inventive and non-alienated relations with technologies lies in 'going into the mould'.

The Technical Object

The example of the moulding of the wet clay brick appears at the start of Simondon's *L'individu* and again at the end of his book *Du Mode*. Thus it provides a kind of common surface through which the two books – one concerned primarily with regimes of individuation, the other with introducing into culture an understanding of the being of technical objects – come into explicit communication with each other; an opportunity not always taken up by Simondon's commentators (who sometimes focus on just one aspect of his philosophy rather than ask about their common ground). In each of the discussions Simondon makes a plea for 'going into the mould'; that is to understand not only the preparations for the metastable state just prior to its resolution through individuation, but also the dynamics of form-taking itself. In *L'individu* the moulding of the clay brick is the only case of a 'technical operation' examined by Simondon (before moving on to other physical and vital individuating systems, and then in the follow-on volume *L'individuation psychique et collective* to the individuation of the group), while *Du Mode* traces numerous examples of technical objects, from car engines and electronic vacuum tubes to the Guimbal turbine. In *L'individu* going into the mould as method opens up a series of examinations of the operations or mediations involved in physical, then vital modes of individuation. In *Du Mode* it is presented rather differently. Simondon argues that the activities of work, those very preparations that create the conditions for the unfolding of the technical operation, also obscure it. Both manager who commands the work, and worker who carries out the commands are outside the technical operation and alienated from it. *Du Mode* presents going into the mould as an image for alternative inventive and non-alienated relations for the worker with the technical object, and develops the notion of technical invention and its potential for the individuation of human collectives – or 'transindividuation' (these ideas are also explored in his text 'Individuation and Invention'[11]). In this section I follow some of Simondon's discussions in *Du Mode* about humans and technical objects and their relations, in order to think more broadly about architecture's relations with the technical aspects of building. Rather than banish materials to the technical outside of architecture proper, could there be better, less alienated relations of knowledge and invention between architects and the materials they build with?

Whereas the technical *operation* is loosely defined in *L'individu*, more by the intention-led chain of operations that create the conditions for its unfolding, than any attempt to set out the category of the technical, the technical *object* is clearly distinguished from other kinds of object in *Du Mode* by the fact of its detachability from the human. A tool, such as a hammer, is not strictly speaking technical, but rather pre-technical because its mobilization requires human action. Following Piaget, Simondon describes a tool such as this as 'deprived of its own individuality because it is grafted onto another individualized organism's body part and because its function is to extend, reinforce, and protect but not replace the latter'. The hammer is merely 'an unfeeling and hardened fist'.[12] And in a different way, an aesthetic object or artwork is not detachable from the human because it is necessarily understood in terms of what it means for humans. In 'Individuation and Invention' Simondon adds that technical invention cannot be identified with a single individual as it can with a poet or a painter.[13] At least as it becomes 'concretized' or more advanced in its development (as opposed to its 'abstract' or 'primitive' state when still made up of 'a convergence of several complete systems'[14]) the technical object has a strange kind of being – because although it is of the human it appears as if natural, able to function independently of the human and also to exist in a way that could not have been imagined all at once by a human:

> The technical object distinguishes itself from the natural being in the sense that it is not part of the world. It intervenes as mediator between man and the world; it is, in this respect, the first detached object.[15]

The concretized technical object has necessarily been through a process that Simondon calls 'evolution' – an iterative process of invention, where the already manufactured object presents possibilities for adaptions through its physical existence in the world. Simondon gives numerous examples of these lineages and supplies the reader with grainy illustrations covered in his scratchy annotations. One example – pictured but not explained in the text – is the telephone. Plate 10 compares a telephone set from 1929 and 1951 according to the caption 'Difference between technical concretization and adaption of the object to the human being: evolution of the telephone'. An arrow points at the column rising out of the dial stand with clips to hold the handset (already a merging of what had been separate mouth and ear pieces in earlier sets) in the 1929 telephone, and contrasts it with the 1951 set where the dial piece is now shaped to include a holder for the handset. Plate 11 shows the 'internal organs of the mobile set' where the more significant technical adaptions occur. The first technical object in a lineage of technical evolution is still a 'primitive object'. It starts out as an 'abstract' technical object; 'the translation into matter of a set of notions and scientific principles that are deeply separate from one another, which are attached only through their consequences and converge for the purpose of the production of a desired effect'.[16] Like the early telephone, they appear as a series of elements each carrying out a particular function as could be pictured by their designer or maker. But gradually new opportunities present themselves – for an element to do more than one job, or to be merged with another element bringing elements together which Simondon terms 'functional synergies'.[17] Through this process – a comingling of the engineer's activity and the object's inherent but as yet unrealized potentials – the technical object becomes over time something more than the original realized diagram of what it was designed to do. Each iteration of the series of objects is something like the face of the crystal that becomes the seed for the next layer.

While Simondon's notion of technical evolution is a key foundation for Bernard Stiegler's philosophy, others are critical of its inherent biologism. Daniela Voss shows that Simondon's formulation draws on his teacher Georges Canguilhem's understanding of technical invention as 'a universal biological phenomenon and no longer simply as an intellectual operation to be carried out by man'.[18] Her concern is the normativity involved here, and Muriel Combes confirms that 'there is indeed in Simondon the idea of normativity to technics'.[19] However Simondon distinguishes his notion of evolution from the purely biological. These are not simply mutations (as in evolution) but 'mutations which are oriented'.[20] Moreover, Simondon's point is surely in the other direction that is to challenge a purely human, or intellectual notion of technical development. By suggesting a biologism to some lineages of technical development, Simondon seeks to undo the notion that technical invention emerges fully formed in an Archimedean moment, direct from the mind of the inventor (then it would still be an abstract object). This iterative process of invention eventually results in an object that is so far from its abstract ancestor (when the relation to the human designer was still evident, and might need to be maintained in an artificial environment that supported it, such as a lab) that it can appear as if not human, or 'concrete' and as if part of nature:

> The technical object, which thought and constructed by man, is not limited to creating a mediation between man and nature; it is a stable mixture of the human and the natural, it contains human and natural aspects; it gives

its human content a structure comparable to that of natural objects and allows for the integration of this human reality in the world of natural causes and effects.[21]

The effects of this detachability from the human are multiple. On the one hand, detachability can have negative repercussions. The machine might seem to challenge humans in so far as it takes over their activity, or appears as a threatening stranger in the case of the robot, imagined by culture that does not embrace technics, as 'animated by hostile *intentions* toward man' or as representing 'a permanent danger of aggression and insurrection against him'.[22] Just as internal synergies developed between its elements, the concretized technical object comes into association with other technical objects as in the case of the electricity grid or the telephone network. Operating at a much greater scale, even forming the environments we inhabit, these 'technical ensembles' (Combes connects this term to the more familiar 'networks') are also alienating if the human is estranged from them. For Simondon there is nothing necessary about this alienation. We could develop technical ensembles in which the human participates, or machines where the technician does more than simply operate them, or act as a cog in their wheel. But to do that we must first embrace technics, technical understanding and include technical objects as part of culture rather than banishing them 'into a structureless world of things that have no signification but only a use, a utility function'.

On the other hand, there are positive implications. The detachability of the concretized technical object means that someone else can pick it up; 'The machine has a sort of impersonality which allows it to become an instrument for another man.'[23] It has the extraordinary characteristic of being both of the human and, because detached and self-sufficient, of nature. As Paolo Virno puts it, Simondon sees the potential of technical objects to, 'give an external appearance to what is collective, to what is species specific in human thought'.[24] But because the concretized object is not just a realization of something imagined, rather the outcome of a process of evolution via a series of objects, it is more than a simple reflection of human thought. These technical objects present themselves as analogous to 'spontaneously produced objects' (not just as representations of thoughts already thought) so that 'one can legitimately consider them as one would natural objects [. . .] one can submit them to inductive study'.[25] Moreover they present as a novel kind of nature that might operate according to different laws and demand or open up new kinds of concepts.

In circumstances where through 'technical activity' we are able to have integrated relationships with such technical objects (and develop a truly technical culture), they offer, Simondon argues, a remarkable potential for collective life. First, technical activity provides a two-way connection between the human and nature:

> In edifying the world of technical objects and by generalizing the objective mediation between man and nature, technical activity re-attaches man to nature through a far richer and better defined link that that of the specific reaction of collective work. A convertibility of the human into the natural and of the natural into the human establishes itself through the technical schematism.[26]

And second, because the technical object is detached from the individual but nevertheless carries something of the human, it has the potential to be one of the means through which the collective finds its common ground and comes into being or individuates. 'The technical world' claims Simondon, 'is a world of the collective which is adequately thought neither on the basis of the brute social [fact], nor on the basis of the psyche'.[27] We might recall here the travellers who encounter the boulder blocking their route. Their resolution of the problem involves mental effort and bodily exertions that enable them to form a system in which each traveller is able to act on the rock. The boulder acts like the crystal seed, discovering a rock-pushing potential in each of them that can only be actualized in their coming together through their common task. And at the same time, at least while they are pushing, they stop being a group of disparate individuals and a rock-pushing collective is born. 'Technical activity' writes Simondon, 'is the model of the collective relationship . . . it is not the only mode and the only content of the collective, but it is of the collective and in certain cases, it is around technical activity that the collective group can arise'.[28] He names this collective form of individuation 'transindividual'.

In the conclusion to *Du Mode* Simondon finally locates *work* as the cause of culture's failure to embrace technics and their potential of technical activity as means for transindividuation:

> To this day, the reality of the technical object has been relegated to the background behind the reality of human work [labour]. The technical object has been apprehended through human work, thought and judged as instrument, adjuvant, or product of work. However, one ought to be capable, in favour of man himself, to carry out a reversal that would enable what is human in the technical object to appear directly without passing through the relation of work.[29]

Malaspina and Rogove use 'work' to translate 'travail' in the original French. But given Simondon's explicit engagement with Marx in this closing chapter, it must also have the sense of 'labour' throughout. Indeed, in 'Individuation and Invention', where Simondon again distinguishes labour from technical activity, Taylor Adkins translates 'travail' with 'labour'.[30] It is through man's labour, Simondon tells us in *Du Mode*, when he offers 'his organism as a tool bearer' that man 'actualizes the mediation between the human species and nature within himself'.[31] Rather than give over to the technical object to 'fulfil this function of relation' man carries it out himself 'through his body, his thought, his action',[32] and in so doing, with his eye on the goal or output of his labour, he fails to see the technical operation itself and its significance as mediation between human and nature:

> The activity of work [labour] is what forms the link between natural matter and form, which comes from man. Work is an activity that succeeds in making two realities as heterogeneous as matter and form coincide and renders them synergetic. And the activity of work makes man aware of the two terms he synthetically relates, because the worker must have his eyes fixed on these two terms which he must bring closer together (this is the norm for work), not on the interiority itself of the complex operation through which this bringing together is obtained. Work masks the relation in favor of the terms.[33]

This is as true of 'the man who orders work to be done' as the worker.[34] Simondon describes him much like most architects, as concerned with the content and with the simplified notion of 'the raw material that is the condition of execution' rather than with the operations that enable taking form to take place. Simondon argues that it is this notion of the technical operation – reduced to work – that has become so embedded in philosophy via the hylomorphic schema:

> The *obscure central zone* characteristic of work has transferred itself to the utilization of the machine: it is now the functioning of the machine: it is not the functioning of the machine, the provenance of the machine, the signification of what the machine does and the way in which it is made that is the obscure zone. The primitive central obscurity of the hylomorphic schema is preserved: man knows what goes into the machine and what comes out, but not what happens in it: an operation takes place in the very presence of the worker in which he does not participate, not even if he commands or serves it.[35]

He describes again the labours involved in the chain of preliminary operations that prepare clay and mould for form-taking, and reminds us that 'the working man prepares the mediation, but he doesn't fulfil it'.[36] Repeating his injunction in the Form and Matter chapter in *L'individu*, he calls for the working man to go into the mould 'with the clay, to be both mold and clay at once, to live and feel their common operation in order to be able to think the process of taking form in itself'.[37] But in *Du Mode* it is less with the aim to understand the system of individuation, than to enable an inventive engagement with the technical operation.

Simondon argues that the technical operation is just as obscured when production is carried out with a tool (when the energy of the human and their form intention is needed) even if the brick-maker has the sense they are accomplishing it, as when it is carried out by a machine (which merely functions). We recall Barthes' machine producing green dressing room tidies here. It is just that when a machine or technical object carries out the task the estrangement from the technical operation becomes much more obvious; the human contribution that came from the worker in the pre-technical production now resides in the object. As Simondon puts it, 'when the technical object is concretized, the mixture of man and nature is constituted at the level of this object'.[38] Although workers could delude themselves that they directly effected communication with nature through moulding the brick, it would be impossible to convince themselves of the same when operating a machine.

Simondon is at pains to point out in this chapter that Marx's account of alienation is insufficient, even if 'it is correct to say that the economic conditions and amplify this alienation' when men do not own the machines they operate.[39] Alienation cannot be addressed solely by change in the means of production, or in Simondon's reduction of Marx's argument, 'to possess a machine is not to know it'. For Simondon, the impoverished relations between humans and technical objects that bring about alienation, are not specifically the result of capitalism or any other mode of production. Indeed in 'Individuation and Invention' he argues that it is only because a civilization becomes attached to productivity (as that which resists and expels creative technical initiations and 'enslaves man and machine') 'that it becomes a civilization of money when certain circumstances turn this mode of exchange into the concrete criterion of productivity'.[40] Change demands that we incorporate technical understanding into culture (Simondon has proposals about school education as well as professional education) and on the degree to which the technical objects we make allow for inventive relations with and through them. His challenge to Marx frustrates some commentators. While at the same time providing an excellent summary of his ideas about technical invention Daniela Voss in particular sees Simondon's bracketing off of social relations and economic and political conditions as a purification that finally leaves him unable to provide a coherent account of the forces that shape technical development.[41] She reminds us that in Marx's analysis the development of new machinery aimed either to enhance productivity or (citing Marx in *Capital*

1) 'for the sole purpose of providing capital with weapons against working class revolt'. Certainly, these are limitations in Simondon's project, and Toscano (whose argument Voss agrees with) is right that Simondon's work does not provide a means to think the 'external capture' of technics by capitalism. However, Simondon brackets off social relations because he seeks to demonstrate a different kind of cause of alienation, by developing as an alternative to either tool or commodity the category of the technical object with its opportunities for thought, culture and the formation of collectives. For Simondon, as Combes puts it, 'only a definitive departure from the paradigm of labour can permit humans to transform their inadequate relation to technics, to nature, and to one another'.[42]

According to Simondon, 'the relation of the worker to the machine is inadequate, because the worker operates on the machine without his gesture continuing the activity of invention in this gesture'.[43] The overcoming of technical alienation therefore requires the fostering of inventive relations with technics, where invention is the engagement with mediation, and the prolonging of individuation. First, education in the history of technics and the development of a new discipline of 'mechanology' is needed to develop understanding of the technical operation. 'For information to be exchanged' meaningfully between man and machine during the technical operation, says Simondon, 'man must possess within himself a technical culture'.[44] Second, machines must be developed that can allow alternative relations with their operators. Simondon points out that manufacturers tend to seal off the internal mechanisms of the machine from the operator, even using warranties to keep maintenance and repair in their own hands.[45] He accepts that economic factors are at work here, but argues that it is only by making machines that have the capacity to be regulated and modulated by their operators in an iterative process 'where man encounters man not as a the member of a class but as a being who expresses himself within the technical object', that alienation can be avoided.

As Alberto Toscano has argued in his essay 'Technical Culture and the Limits of Interaction: A Note on Simondon', Simondon differentiates alienating labour from a better kind of technical activity[46] which instead 'involves not only the utilization of the machine, but also a certain coefficient of attention to technical functioning, maintenance, regulation, betterment of the machine, which prolongs the activity of invention and construction'.[47] Simondon envisages a transformation of work in which we would neither direct the technical operation from outside it nor become a cog in the machine but would become part of the technical ensemble's functioning in an iterative and inventive capacity. The ethical question is not one of the operator governing or dominating the machine, or vice versa. For Simondon 'invention' does not occur in an ideal realm to be realized in matter and imposed upon it, but comes about through the transaction of operator and machine, as part of an individuating system.

Inventive Relations

If Chapter 6 took up the system of individuation as it unfolds in *L'individu* to establish a theory of building materials in which it is the mediations they make possible that are decisive, Simondon's discussion at the close of *Du Mode* prompts us to consider the relation of the architect (as 'manager') to building materials (as technical objects) and to technical culture more widely. It also reminds us that the architect's 'commands' have a bearing on the building worker[48] – whether the fabricator who fixes a performance-engineered curtain walling system or the labourer who mixes and casts concrete on site and, at a further remove on those who produce building materials. The wide range of brick-making methods around the world documented by film-maker Harun Farocki in his epic film *In Comparison* (2009) connects Simondon's discussion to the building materials industry. Farocki shows us a woman labourer in the hazy heat of India's Hinjawadi brickfields packing wet clay into a metal double mould, using methods as simple as those described in Simondon's Form and Matter chapter. In another sequence, Farocki captures the boredom and alienation of automated production in a facility in Olfen-Vinum, Germany where a single operative sits in front of a few switches, while machines hum and clank around him, assembling fully formed brick panels.[49] The effects of architects' design choices on the people who build with and produce materials are important, in Marx's terms as well as in Simondon's. But built into the structure of the architectural profession is the bracketing of design off from production. Architects need to be more aware of the implications of their practices on how labour is organized. These crucial issues are addressed in the work of Sérgio Ferro, Christine Wall, Pedro Fiori Arantes and Linda Clark among others, but are beyond the scope of this book.[50]

In *Du Mode* Simondon makes a case for a culture of technics that is particularly pertinent to architecture and to our study of building materials. He argues that it is precisely by considering technical objects in terms of utility, and as a means to an end that we fail to develop an understanding of them or to develop inventive relations with them. The hylomorphic schema, which is also blind to the technical operation, contributes to and sustains this failure. These arguments can equally be applied to architecture's approach to building materials. Materials are relegated to architecture's technical underside. They are absent in the valorized drawing, and left to the specification – a 'supplementary' written document, which these days only technical experts can understand. Few architects know much about how the materials they select are developed or made ready for building. Even the dominant form of specification today – performance specification – makes the use a material will be put to its defining characteristic, as if materials had no other existence than their utility; no qualities; no genesis; even no name. Just as Simondon describes, technical operations are veiled. Like Barthes' dressing room tidies, cast concrete takes on the architect's drawn form out of sight as if by magic. The operational films and interlayers in performance-engineered glass hide themselves in plain sight, imperceptible and transparent. Moreover, the vast corporate complex with its chains of processes, material and immaterial, that produce and sediments them in contemporary building culture is just as unseen and unknown. Here I agree with Voss, when she turns to Gille in his *History of Techniques* to argue that because today technical invention usually demands extensive funding, scientific knowledge and lab capacity, we need to 'study the complex institutional contexts that foster fundamental research and direct the process of invention' and the techniques that drive development such as patents and subsidy.[51]

First, then, Simondon's injunction to 'go into the mould' is an imperative to know the technical operation. A pre-requisite would be at least that the processes of individuation, of mediation and becoming compatible, could be known and not (to use Muriel Combes' term) 'black-boxed' as they are in the terms of the hylomorphic schema. The technics of building and building materials must become part of architectural culture. Second, 'going into the mould' demands that we take up an inventive relation with building materials. Unleashing the transindividual potential of a better relation with technics and, as Combes puts it, 'reducing alienation means showing that technical objects are not the Other of the human, but themselves contain something of the human: the "object that comes of technical invention carries with it something of the being that produced it"'.[52]

Simondon gives us two ways to consider invention with respect to the technical object. The first, his focus on more inventive relations between technician and machine, is only directly relevant to our study of building materials in a few rare cases where architects work directly with materials through iterative processes. Chandler's Wall One (introduced in Chapter 4) is one example. There an adjustable rig held the fabric formwork mould in place. The team of fabricators could tighten and release the fabric bag as the concrete was being poured and setting. Together with rig, bag and sluggish concrete, the team participated in the dynamic operations of the wall's taking form.[53] At the Hub for Biotechnology in the Built Environment (HBBE) at my own work place, Newcastle University, architectural researchers work alongside biologists and bacteria, growing living materials and experimenting with coaxing useful forms and behaviours from them.[54] However, following Simondon's notion of technical invention means more than architects becoming involved in the fabrication process. For example, where architects are embracing 3D printer technologies that allow them to have a much closer relation to manufacture, the technical operation could not be more 'black-boxed'. At first sight, this technical system replicates the hylomorphic schema more precisely than either Barthes or Aristotle could ever have dreamed possible. This technology appears, at least at first glance, as geographer Tom Roberts has put it with reference to Simondon, 'to merely reinforce the metaphysical distinction between acting human subject and passive material substrate', and provides perhaps a more readily graspable exemplar of hylomorphism than the wet clay brick.[55] But as Roberts suggests with reference to Michael Hansmeyer and Benjamin Dillenburger's *Digital Grotesque* project, but can perhaps be better seen in Helena Westerlind's larger scale experiments with concrete deposition, 3D printing can allow for iterative and inventive relations between technology, fabricator and material.[56] Flavio Bevilacqua also finds in digital manufacture the possibilities of non-hylomorphic design processes in line with Simondon's thought.[57] Indeed, for Stefano Mazzilli-Daeschel the kinds of community workshops that have been able to emerge due to the proliferation of such digital technologies – makerspaces, FabLabs and hackerspaces – can be understood in terms of what Andrea Bardin names 'Simondon's pedagogical-political project', as fostering technical culture.[58]

More directly applicable to the case of building materials are Simondon's ideas about concretization and technical evolution. Concretization is a process through which each iteration of the object's development is open to further adaption. Each iteration is an individuation between engineer and the possibilities offered by the object. If the inventive relation between an operative and a machine would occur as a to-ing and fro-ing across a few hours, the inventive relation in concretization unfolds over years or decades, each iteration holding traces of the last design interaction between object and designer, as it becomes the new nature (or surface, to recall the crystal) for the latest designer. Despite her reservations about Simondon's bracketing off of economic and social considerations, Voss explains this beautifully:

> Simondon describes the technical invention as a process that surpasses the individual psyche. First of all, in most cases, there is not one inventor but a succession of inventors that, if separated by time and space, communicate through already existing technical objects. These technical objects, detachable as they are from the space and time of their creation, support what Simondon will call relations of cumulative participation.[59]

Can we imagine what this kind of inventive relation could mean for architects with respect to building materials? Certainly, it could include the practices mentioned above, and others we have seen earlier in the book such as DSDHA's flame treatment of timber or SWA's experiments with straw bales (both discussed in Chapter 6). A pavilion for the park next to Tower Bridge, an architect's own house and office; these are both small, bespoke, one-off projects where material creativity is possible. But we should be wary of only considering this in terms of a return to artisanal practices. Certainly, this is not what Simondon advocates. As Thomas LeMarre explains, 'he is not interested in a return to a premodern guild or artisan formation in which the role of humans was closer to the technical individual.'[60] In fact Simondon identifies the evolved technical objects and ensembles that have the potential to offer man a connection to nature 'much richer and better defined than that of the specific reaction of collective work' as emerging in the 19th century with industrialization, and draws his examples from the vast factories and technical networks that existed in the late 1950s when he wrote *Du Mode*. Most contemporary building makes use of materials produced off-site and at large scale. They are mass-produced and commonplace. They make up the ordinary fabric of our everyday built environments, but as I have started to show here, there are transformations in their ontogenesis and capacities that go unseen and unremarked. To leave these operations black-boxed, to consider them preliminary to architecture's practices and outside them, is to forego the possibility of being both critically aware of them, and of invention or 'cumulative participation' with them.

Might there be architects who operate rather more like the technicians and engineers Simondon envisages, working with manufacturers on the development of building materials, or indeed a new 'mechanology of architecture' as the architect and theorist João Marcos de Almeida Lopes proposes?[61] In a Marxist reading of Simondon (and Ferro) Lopes locates a possibility to understand architecture in terms of the technical ensemble. Indeed Simondon makes a direct reference in *Du Mode* to the building site as a temporary, yet 'highly developed and complex' technical ensemble.[62] For Lopes, approaching architecture as individuation allows us to see it as bringing into communication 'a vast collection of material, operative, cultural and intellectual circumstances coming from different areas of knowledge'.[63] Thus a single disciplinary understanding is inadequate. In particular, architecture (and here he refers to Brazilian historiography, but it is as applicable to the wider Western traditions) 'not only neglects the place of fabrication of its objects, but it also blocks any possibility of taking fabrication as a core issue of the debate about how architecture is produced'.[64] Following Simondon, Lopes identifies this position with the hylomorphic schema, as we have also argued. He also proposes that Simondon shares a sensibility to technology with Marx, and cites a short section from *Capital Vol. 1*:

Technology discloses man's mode of dealing with Nature, the process of production by which he sustains his life, and thereby also lays bare the mode of his social relations, and of the mental conceptions that flow from them.[65]

Lopes reads Simondon's use of 'travail' in the concluding chapter of *Du Mode* as referring not to work in general, but to heteronomous labour. He argues that only under conditions in which labour is the free production of man's material life (he uses the term 'free labour' following Ferro and William Morris), can technical activity be a means to a non-hylomorphic relation with nature. That Lopes turns to Simondon at all, suggest that he too considers the kind of technical object or activity to be determining.

In this theoretical text, Lopes does not mention his own work with the São Paulo-based technical aid centre USINA CTAH (Centro de Trabalhos para o Ambiente Habitado, or FACTORY – Work Centre for Living Space), which he founded with others in 1990. Their projects exemplify the alternative architectural relations he proposes. USINA CTAH work with popular movements to construct community centres, schools and childcare centres, and in particular housing. Often working with groups who have, through sustained political struggle, secured land on which to build their own homes, USINA CTAH enable them to establish a 'mutirão' or 'collective effort' to carry out construction work themselves. Of course, it is economically advantageous for the future residents to build themselves rather than work to pay off a mortgage over many years, but USINA CTAH's projects also use the collective process of building as opportunities for new forms of social relation. Importantly, each of the housing schemes evolves around a technical innovation – a lightweight modular ceramic brick that is easy to build with and offers many permutations, even for projects at considerable scale. For example, Copromo, built in the municipality of Osaco in 1990 provides 1000 residential units. Only the steel stair towers, which were built at the start of the project and used as hoists for materials, were constructed by an external contractor.[66] Lopes explains that the individuation of an architectural technical object takes place though, 'the specificities of the work of the architect, structural engineer, builder and especially of the building-site workers – foremen and clerks, steel fixers and carpenters, bricklayers and labourers'.[67] In the case of USINA CTAH, the architect enters into this process at the level of designing materials fabrication. Altered social relations are achieved through an approach to architectural design that embraces the possibilities of technical invention.

One contribution of this book has been to make use of descriptions in the specification to open up the 'black box' of building materials. This document could itself be a site for invention – developing new languages of material description that could build on and transgress the normative frameworks that have been collected here. Examples are few and far between. Heidi Svenningsen Kajita has developed architectural studio briefs that ask students to subvert languages and concerns of normative specification, and pursues these ideas in her own research project '(Im) Possible Instructions: Inscribing use-value in the architectural design process'.[68] Architectural theorist Brady Burroughs invents audacious, tongue-in-cheek 'Architectural Room Specifications' for the fictitious refurbishment of a row house designed by Aldo Rossi in Mozzo, Italy as part of her development of queer feminist design research. 'Where possible' states one of the instructions in '01 General Information', 'small local businesses, especially those owned by women and "minoritarian subjects" should be given priority for contract of services' and 'everything is to be done with enthusiasm, care and a *love ethic*'.[69] However the paper-based specifications these projects are based on, and those I have studied in this book – from the single parchments from the early 18th century that were never intended to be read by the builder, to the thick bound books of the LCC in the 1960s with their rich prose and intimate details of processes on-site – have largely been superseded by electronic versions for large corporate practices, and due to the cost of the NBS products today, smaller practices are finding other ways to prescribe materials and qualities of work – such as dialogue with builders, adapting the schedule of works or through notes on the drawing. Notably, the NBS promotes its products to two groups; architects, engineers and specifiers on the one hand, and building product manufacturers on the other, who are invited to pay to, 'Market your products through the NBS platform and start winning more orders'.[70] Product placements appear as options for selection with manufactures supplying bespoke clauses. These new forms of specifications are to some extent black-boxed to the architect by frameworks of constantly updated standards and degree of specialist technical knowledge they require; neither intervention nor invention with their techniques seems possible. And for the researcher, they are less open to the kinds of close readings I have been able to make of earlier documents in parallel with Simondon's philosophy – in order to follow systems of material over materials as substance or matter.

This book has also made a case for more architectural theory and history that folds technical and industrial activities into its discourses.[71] Certainly we need to equip ourselves with a critical understanding of some of the more insidious developments taking place at the level of production. And importantly, while these technical operations remain black-boxed, the potential to be inventive with them is also curtailed. Of course, the steps taken here towards a theory of materials as technical systems are themselves only preliminary. The possibilities of technical invention would need to be explored in the concrete realities of architectural practice. To 'go into the mould', to take part in and alter the mobilization of building materials, for more than the expansion of the possibilities of form-making, would require grappling with financial instruments, regulatory frameworks, planning, policy, industry standards, and with materials, their manufacture and material practices.[72] Inserting into architectural discourse a proper consideration of technical activities in all their breadth might at least begin to create the conditions for such activity.

Notes

1 Introduction

1 Cyril Stanley Smith, *A Search for Structure: Selected Essays on Science, Art and History* (Cambridge, Massachusetts: MIT Press, 1982), 115.
2 Ibid., 377–8.
3 Ibid., 378.
4 See Lars Spuybroek, *NOX: Machining Architecture* (London: Thames and Hudson, 2004); Paola Antonelli with Anna Burckhardt, *The Neri Oxman Material Ecology Catalogue* (New York: The Museum of Modern Art, 2020).
5 Jane Bennett, *Vibrant Matter: A political ecology of things* (Durham, N.C.: Duke Press, 2010); Rachel Armstrong, *Vibrant Architecture: Matter as a Codesigner of Living Structures* (Warsaw: De Gruyter Open: 2015).
6 Peg Rawes (ed.) *Relational Architectural Ecologies: Architecture, nature and subjectivity* (London: Routledge, 2013); Hélène Frichot, *Creative Ecologies: Theorizing the Practice of Architecture* (London: Bloomsbury, 2019).
7 Maria Puig de la Bellacasa, *Matters of Care: Speculative Ethics in More Than Human Worlds* (Minneapolis: University of Minnesota Press, 2017); Astrida Neimanis, *Bodies of Water: Posthuman Feminist Phenomenology* (London: Bloomsbury, 2017).
8 For Julieanna Preston's works see http://www.julieannapreston.space [accessed 20 March 2020]. Those referred to here are *murmur* (2017); *A Reconciliation of Carboniferous Accretions* (2014); *water-logged* (2016) and *BALE* (2011).
9 For Preston's projects with building products see for example SHEAR (2008) and SWELL (2008); described in Julieanna Preston, *Performing Matter: Interior Surface and Feminist Actions* (Baunach: Spurbuchverlag, 2014), especially 37–74.
10 Walter Gropius, cited in Annemarie Jaeggi, *Fagus: Industrial Culture from Werkbund to Bauhaus* (New York: Princeton Architectural Press, 2000), 47.
11 This discussion is made in Sérgio Ferro, *dessin/chantier* (Grenoble: Éditions de la Villette, 2005); Sérgio Ferro, *O canteiro e o desenho* (São Paulo: Projeto Editores, 1979). For a shorter introduction to Ferro's argument in English see, Sérgio Ferro, trans. Ricardo Agarez and Silke Kapp, 'Dessin/Chantier: an introduction' in Katie Lloyd Thomas, Tilo Amhoff and Nick Beech (eds), *Industries of Architecture* (London: Routledge, 2015).

12 For this argument, see Silke Kapp, Katie Lloyd Thomas and João Marcos de Almeida Lopes, 'How to Look at Architecture from Below', an introduction to Sérgio Ferro, 'Concrete as Weapon', in *Harvard Design Guide*, No.46 – F/W 2018, 161.

13 Mark Hayward and Bernard Dionysius Geoghegan 'Introduction: Catching Up With Simondon', *SubStance* 41, No 3, 2012 (Issue 129), 3–15, 4.

14 For example, the conference 'Architecture and Bureaucracy: Entangled sites of knowledge production and exchange' in Brussels, 30–31 October 2019. www.architectureandbureaucracy.be (accessed 7 June 2020). Recent research on technical literatures includes Ricardo Agarez's study of *memória descritiva* or architect's official design rationales, in his *Algarve Building: Modernism, Regionalism and Architecture in the South of Portugal, 1925–1965* (London: Routledge, 2016) and Michael Osman's 'Regulation through Paperwork in Architectural Practice' in *Modernism's Invisible Hand: Regulation and Architecture in America* (Minneapolis: University of Minnesota Press, 2018), 165–89.

15 For example, Articles of Agreement for a town house for Sir William Heathcote at St James Square, London. Architect – Henry Flitcroft (1734–6). RIBA Archives, HeW/1/1/2.

16 Specification of Works for a smoked sausage manufactory, 5 Fairhazel Gardens, NW6. Architect – Gerhard Rosenberg (1934–5). RIBA Archives, SaG/17/7.

17 For example, Specification for the Works at the Elfrida Rathbone School for the Educationally Subnormal, Camberwell. Architect – John Bancroft with London County Council (1961), RIBA Archives, LCC/AD/1.

18 Architectural historians working in this area include Inge Bertels, Merlijn Hurx, Jeroen Cornilly and Gabri van Tussenbroek (on the Low Countries context – Belgium and the Netherlands), Pierre Edouard Latouche and Jessica Garcia Fritz (on the North America context), Sarah Melsens (on Pune, India), and Tilo Amhoff (on the 18th and 19th centuries in the UK).

19 John Summerson, *Georgian London*, (London: Pleiades Books, 1945); Howard Davis, *The Culture of Building* (Oxford: Oxford University Press, 2006).

20 Merlijn Hurx, *Architecture as profession: The origins of architectural practice in the Low Countries in the fifteenth century* (Brussels: Brepols Publishers, 2017).

21 Michael Osman, 'Specifying: The Generality of Clerical Labor' in (eds) Zeynep Çelik Alexander and John May, *Design Technics: Archaeologies of architectural practice* (Minneapolis: University of Minnesota Press, 2020), 129–62.

22 Mhairi McVicar, *Precision in Architecture: Certainty, Ambiguity and Deviation* (London: Routledge, 2019).

23 Michael Ball, *Rebuilding Construction* (London: Routledge, 1988), 37.

24 For an exploration of the language of specification in relation to Jacques Derrida's *Glas*, see Katie Lloyd Thomas, 'Specifications: Writing Materials in architecture and philosophy' in *ARQ* 8, 3/4 (Dec 2004), 277–83, and in relation to the poetry of Francis Ponge, see Katie Lloyd Thomas, 'Specifying

Materials: language, matter and the conspiracy of muteness' in Marco Frascari, Jonathan Hale and Bradley Starkey (eds), *From Models to Drawings* (London: Routledge, 2007), 242–52.

25 'Countless scholars have meticulously traced the emergence of the modern concept of form from Kantian aesthetics to twentieth century modernism, but what if the concept's salience in modernist discourses also had to do with more mundane – what might perhaps be called "technical" – concerns?', Zeynep Çelik Alexander, 'Scanning: a technical history of form', in Zeynep Çelik Alexander and John May (eds), *Design Technics: Archaeologies of architectural practice* (Minneapolis: University of Minnesota Press, 2020), 71.

26 Gilbert Simondon, *Du mode d'existence des object techniques* (Paris: Aubier, 2001); Gilbert Simondon, trans. Cécile Malaspina and John Rogove, *On the Existence of Technical Objects* (Minneapolis: Univocal Publishing, 2017). An English translation by Ninian Mellamphy (University of Western Ontario, 1980) of Part One was freely available online prior to 2017 and used in the preparation of this research. See http://www.rybn.org/ANTI/ADMXI/documentation/ADMXI/II._ALGORITHM_ENGINEERING/1958/Simondon_On_the_Mode_of_Existence_of_Technical_Objects.pdf (accessed 7 June 2020).

27 Gilbert Simondon, *L'individuation à la lumière des notions de forme et d'information* (Grenoble: Editions Jérôme Millon, 2006).

28 Gilbert Simondon, trans. Taylor Adkins, *Individuation in Light of Notions of Form and Information* (Minneapolis: University of Minnesota Press, 2020), and Gilbert Simondon, *Individuation in Light of Notions of Form and Information, Volume II: Supplemental Texts* (Minneapolis: University of Minnesota Press, 2020).

29 Gilbert Simondon, *L'individu et sa genèse physico-biologique* (Paris: Presses Universitaires de France, 1964). An English translation of the introduction to *L'individu* appears as 'The Genesis of the Individual', trans. Mark Cohen and Sanford Kwinter in Sanford Kwinter, Jonathan Crary (eds), *Incorporations* (New York: Zone Books, 1992). Taylor Adkins' English translation of Chapter 1 (parts I and II only) is freely available online as 'The Physico-biological Genesis of the Individual' at http://fractalontology.wordpress.com /2007/ 10/03/ translation-simondon-and-the-physico-biological-genesis-of-the-individual/ (accessed 7 June 2020).

30 Gilbert Simondon, *L'individuation psychique et collective* (Paris: Aubier, 2007).

31 John Protevi, *Political Physics: Deleuze, Derrida and the Body Politic* (London: Athlone Press, 2001).

32 Peter Salter, *TS: Intuition and Process* (London: The Architectural Association, 1989), 48. Susannah Hagan has also noted the advantages for an environmentalist analysis, of Salter's account of materials which recognizes change and modification, Susannah Hagan, *Taking Shape: A new contract between architecture and nature*, (Oxford: Architectural Press, 2001), 90.

33 Karl Marx, *Capital Volume 1*, trans. Ben Fowkes (London: Penguin Books, 1990),184, my emphasis.

34 Much of the discourse around the gold and silver standards in use in Marx's day concerned the very problems of maintaining monetary equivalence in the face of major changes in the availability of material and the techniques of production. The 1850s saw the discovery of vast new reservoirs of gold in California, Australia and South America and in fact Engels added a long footnote to the 'Money' section of *Capital* explaining enormous changes to the relative values of silver and gold as 'the result of a revolution of the mode of production of both metals', which included the gold rushes of the time and new techniques of extraction. Ibid., 241.

35 Martin Heidegger, 'The Origin of the Work of Art' in *Poetry, Language, Thought*, trans. Albert Hofstatder (New York: Harper & Row Publishers, 1971), 27.

36 Ibid., 28.

37 Ibid.

38 Specification of works for a 'Gothic' villa at Godden Green, near Sevenoaks, Kent for Mr Usborne. Architect – Joseph Fogerty. RIBA Archives, FoJo/1/1.

39 Simondon, *Du Mode*, 15.

40 Ibid. My emphasis.

41 Ibid., fn.1. As translated in 'The Genesis of the Individual', 318, fn.5. Also see fn.13 for a more direct description of photosynthesis as vital individuation.

42 Alberto Toscano, *The Theatre of Production: Philosophy and Individuation between Kant and Deleuze* (Basingstoke: Palgrave Macmillan, 2006) 139.

43 On the importance of this principle to modernist architecture see Adam Sharr, *Modern Architecture: A very short introduction* (Oxford: Oxford University Press, 2018).

44 Emily Thompson, *The Soundscape of Modernity: Architectural Acoustics and the Culture of Listening in America, 1900–1933* (Cambridge, Massachusetts: MIT Press, 2004).

45 Jean-Hughes Barthélémy, 'Encyclopedism' in 'Fifty Key Terms in the Works of Gilbert Simondon' trans. Arne de Boever in Arne de Boever, Alex Murray, Jon Rogge and Ashley Woodward (eds) *Gilbert Simondon: Being and Technology* (Edinburgh: Edinburgh University Press, 2012), 211.

46 Adrian Mackenzie, *Transductions: Bodies and Machines at Speed* (London: Continuum, 2002) 17.

47 Ibid., 25, fn.3.

48 Simondon, *L'individu*, 21. As translated in 'The Genesis of the Individual', 315.

49 Ibid., 313.

50 Ibid.

51 *L'individu*, 20, as translated in 'The Genesis of the Individual', 314.

52 Ibid., 19. As translated in 'The Genesis of the Individual', 313.

53 Jean Piaget, trans. Marjorie Walden, *Judgment and Reasoning in the Child* (London: Routledge, 2014, first publ. 1928), 233.

54 *L'individu*, 20, as translated in 'The Genesis of the Individual', 314.

55 Ibid., 'Clearly transduction cannot be presented as a logical procedure terminating in a conclusive proof . . . I see it as a mental procedure, or better the course taken by the mind on its journey of discovery. This course would be *to follow the being from the moment of its genesis, to see the genesis* of the thought through to its completion at the same time as the genesis of the object reaches its completion.' As translated in 'The Genesis of the Individual', 314. Jacques Garelli explains that for Simondon transduction cannot be included within strictly logical procedures: 'The situation prevents the conceiving of transduction as a simple logical process, which would have the cognitive function of locating and classifying other strictly logical figures such as deduction or induction.' 'Transduction et information' in *Gilbert Simondon* (Paris: Albin Michel, 1994) 56.

56 Simondon, *L'individu*, 21. As translated in 'The Genesis of the Individual', 315.

57 Ibid.

58 Ibid.

2 Specifying Building Materials

1 Sigfried Giedion, *Mechanization Takes Command: A contribution to anonymous history* (Minneapolis: University of Minnesota Press, 2013; first published 1948), v.

2 See 'Table of Contents Page', *Specification*, Vol. 1, 1898 and 'Scope and Contents,' brochure for *NBS Building* (Newcastle: NBS, 3 Nov 2009).

3 We find bell-hangers and well-sinkers but not asphalters included as trades in the first edition of *Specification*, Vol. 1, 1898. But there is a brief section for Asphalters included in the specification for Valentine Harding's modernist reinforced concrete house which incorporated flat roofs, see Specification of Works for a House at Farnham Common, Bucks, 1934 (RIBA Archives, SaG/17/3).

4 Andrew Knightly Brown, 'Stone (Extracting Blood From)' on *The Spec Man Blog*, 13 August 2012, www.andrewkb.net/blog/20120813.htm2013 (accessed 2 November 2013).

5 Although present-day specifications in the UK take many forms, the use of this A–Z classification of work sections – UNICLASS – has been commonly applied since its introduction in 1997.

6 Michel Foucault, *The Archaeology of Knowledge* trans. A. M. Sheridan Smith (London: Routledge, 1972) 57.

7 Ibid., 103.

8 As Nick Schumann, one of the original founders of Schumann Smith explains, in some global contexts, the requirement to avoid naming particular materials is in fact enshrined in regulation: 'European Regulations for the procurement of public buildings does not allow the naming of products and in many other parts of the world clients insist on listing at least three products and manufacturers for every element, which for natural materials or those of particular aesthetic appearance creates quite a challenge for the designer.'

Nick Schumann, 'The Watchful Protector of Quality', *Architects Journal*, 28 October 2013, http://schumannconsultltd.com/blog/watchful-protector-quality-architects-journal-october-column/#sthash.1IkYqEIE.dpbs (accessed 2 November 2013).

9   Howard Davis, *The Culture of Building*, (Oxford: Oxford University Press, 2006) 186 and 193.

10 Specification for Newgate Gaol and the Sessions House at the Old Bailey, London (1769), architect – George Dance the Younger. RIBA Library EW No. 790.

11 Particulars for a Bakehouse, dwelling and lofts in parish of St Anne, Middlesex (1770), contractor – Robert Wilson, client – John Steinmetz. RIBA Archives WiR/1.

12 Specification of Works for a smoked sausage manufactory, 5 Fairhazel Gardens, NW6 (1934–5), architect – Gerhard Rosenberg, RIBA Archives, SaG/17/7.

13 Tony Allot, 'NBS; A Progress Report' in *RIBA Journal*, February 1971, 82.

14 Foucault, *The Archaeology of Knowledge*, 98.

15 Ibid., 123.

16 John Summerson, *Georgian London*, (London: Pleiades Books, 1945).

17 Articles of Agreement for a town house for Sir William Heathcote at St James Square (1734–6), RIBA Archives, HeW/1/1/2.

18 Articles of Agreement for a town house for Sir William Heathcote (1734–6).

19 An illustration of this specification is included in Davis, *The Culture of Building*, Figure 8.4, 186.

20 Particulars for a bakehouse, dwelling and lofts (1770).

21 Specification for Newgate Gaol and the Sessions House at the Old Bailey, London (1769).

22 Dorothy Stroud, *George Dance, Architect 1745–18 – 1825* (London: Faber and Faber, 1971). The building incurred huge debts and in 1778 a special act of parliament was passed to allow money from the Orphan's Fund to be cover the remaining costs.

23 Summerson, *Georgian London*, 19.

24 I am grateful to Tilo Amhoff for drawing my attention to his research on contractual documents in the remarkably complete John Soane archive. For more on this see Katie Lloyd Thomas and Tilo Amhoff, 'Writing Work: Changing Practices of Architectural Specification' in Peggy Deamer (ed.) *The Architect as Worker: Immaterial labor, the creative class, and the politics of design* (London: Bloomsbury, 2015).

25 The particulars and estimates of the several works for Tendring Hall (1784), Soane Museum, SM 28/3/1a/1–6.

26 Tilo Amhoff, *Adapting the architect's products to the capitalist building production: The development of the Legal Obligations, Building Specifications and Working Drawings in the first half of the nineteenth-century in England*, unpublished dissertation submitted for the MSc in Architectural History, The Bartlett, University College London, 2003/2004, 23.

27 Jack Bowyer, *A History of Building* (Eastbourne: Orion Books, 1983) 198.

28 John Gelder, *Specifying Architecture: A Guide to Professional Practice* (Milsons Point: Construction Information Systems Australia, 2001) 15.

29 Ibid., 68.

30 Ibid., 8.

31 For an excellent analysis of the impact of gross tendering on the practice of specification see Amhoff, *Adapting the architect's products*.

32 Bowyer, *A History of Building*, 239.

33 Specification for Newgate Gaol and the Sessions House at the Old Bailey (1769), 'A Description of the Painter's Work', (RIBA Library, EW No. 789).

34 Thomas Donaldson, *Handbook of Specifications* (London: Lockwood and Co., 1860).

35 The RIBA was established in 1834. No date is recorded for Mocatta's donation but the volume includes a specification for the *London Fever Hospital* (still standing and converted into flats) which was not designed until the late 1840s. David Mocatta, *Specifications* circa 1833–43 (RIBA Archive, MoD/1/1).

36 Alfred Bartholomew, *Specifications for Practical Architecture* (London: John Williams & Co., 1840).

37 Thomas Leverton Donaldson, *Handbook of Specifications: or, practical guide to the Architect, Engineer, Surveyor, and Builder, in drawing up specifications and contracts for works and constructions* (London: Atchley and Co., 1859).

38 *Specification*, Vol 1, 1898, 6.

39 Ibid.

40 Ibid., 5.

41 Frank Macey, *Specifications in Detail* (London: E. & F.N. Spon, 1898). This book was re-edited and published throughout the 20th century and the 1904 edition has been reprinted, *Specifications in Detail*, (Shaftesbury: Donhead, 2009).

42 See letter from Miss Sylvia Locke, RIBA Technical Section, to Roderick + Innes and Co. Ltd., chartered surveyors, 16 April 1964, (RIBA Library, Specifications Panel (of the technical information committee) Papers Box 1): 'A great many architects have developed standard specification clauses for their architecture offices and so have some local authorities such as the LCC and the Yorkshire Development Group for Housing. There is the annual publication of the Architectural Press 'Specification' which gives . . . examples of appropriate specification clauses.'

43 See for example, The Specification for the Works at the Elfrida Rathbone School for the Educationally Subnormal, London 1961, architects – London County Council (LCC) with John Bancroft, project architect. (RIBA Archives, LCC/AD/1) which will be explored in detail in later chapters.

44 Harold J. Rosen refers to this cut and paste technique when he describes 'the specification writer of yesterday with glue pot and scissors' (who will be replaced by the materials scientist) in 'The Future of Specifications Writing – Part II' in *Progressive Architecture*, September 1969, 206.

45 For example see the Specification for the Erection of a School house at Hammersmith (1819) architect – Charles Fowler, (RIBA Archives, FOC/1/1).
46 For examples see Specification for Shops/Offices, Oxford St./Rathbone Pl. 1909, architects – Holden and Adams. (RIBA Archives, AHP/1/18/1) and Specification of Works for a House at Farnham Common (1934).
47 John Carter, 'National Building Specification', *The Architects' Journal Information Library*, 19 March 1969, 759.
48 Michelmore's Report, Panel meeting in 1963, RIBA Library, Specifications Panel Papers Box 1.
49 Allen Ray-Jones report, Panel meeting notes, 1963 (RIBA Library, Specifications Panel Papers Box 1).
50 Maurice Golding (assistant secretary to the RIBA technical section) in notes to Anthony Laing, chair of the Specifications Panel, 9 July 1964 (RIBA Library, Specifications Panel Papers, Box 1).
51 Letter from Anthony Laing to Maurice Golding, 20 October 1964 (RIBA Library, Specifications Panel Papers Box 1).
52 Report by John Carter on RIBA proposal for the NBS, November 1968 (RIBA Library, RIBA Services Working Group Papers, Box 1).
53 Mr George report, Panel meeting notes, 1963 (RIBA Library, Specifications Panel Papers, Box 1).
54 Tony Allot, 'NBS; A Progress Report' in *RIBA Journal*, February 1971, 82.
55 *Construction Indexing Manual* (London: RIBA Publications 1969).
56 See L.M. Giertz, *SfB and Its Development 1950–80* (Dublin: CIB/SfB International Bureau, 1982) 3.
57 *National Building Specification*, (London: RIBA Publications Ltd., 1973)
58 Ibid., 13.
59 Colin McGregor, email correspondence with Katie Lloyd Thomas, 6 March 2006.
60 Michelmore, Report, 1963.
61 These categories are described in full in Gelder, *Specifying Architecture*, 121–41.
62 Specification for the H. . .y Building, (2006).
63 Tony Brett at first explained this as a difference between natural and manufactured materials. Interview with Katie Lloyd Thomas, 1 November 2007.

3 Naming Materials

1 The term 'gopher' has been translated in some versions of the Bible as planed or squared timber but may also have referred to a species of wood such as cedar, cypress, reed or boxwood.
2 *The Holy Bible*, Revised Standard Version (New York: Collins, 1971) 5. This passage is referred to as a specification by Gelder, *Specifying Architecture*, 189.

3 Foucault, *The Archaeology of Knowledge*, 47–8.
4 Specification for Newgate Gaol and the Sessions House at the Old Bailey (1769), 'A Description of the Carpenter's Work to be Done in Building the Sessions-House in the Old Bailey'.
5 Specification for two houses on Marylebone Street in Mocatta, *Specifications* (1833–43). Memel is the German name for Klaipėda, a port town of current day Lithuania; Dantzic is Gdansk and Christiana is current day Oslo.
6 For an account of the timber trade see Sing C. Chew, *Logs for Capital: The Timber Industry and Capitalist Enterprise in the 19th Century* (Westport: Greenwood Press, 1992).
7 Specification of Works for a House at Farnham Common (1934).
8 Specification for Sure Start, St Anne's Ward, Colchester 2004, architect – DSDHA, Section J42 Single Layer Roofing.
9 ThermoWood® is the name for wood which has undergone a treatment process patented by the Finnish ThermoWood Association. Both hardwoods and softwoods can undergo this process so a variety of products come under this name. See brochure at www.alternativewood.ca/brochures/English/Thermowood%20Summary%20Brochure.pdf (accessed 18 February 2010), DSDHA Specification for Sure Start (2004), Section H21 115.
10 Specification for Sure Start (2004), Section H21 115.
11 Victoria Ballard-Bell with Patrick Rand, *Materials for Architectural Design* (London: Laurence King Publishing, 2006).
12 Arthur Lyons, *Materials for Architects & Builders* (Oxford: Elsevier Ltd., 2010).
13 F.R.S. Yorke, 'Preface' in *Specification*, 1935, iii.
14 'Olympia, 1936' in *The Builder*, 4 September 1936, 296.
15 Specification of work for the London Fever Hospital 1833–44, Mocatta, *Specifications*.
16 Specification of sundry works required in rebuilding the house no.1 Princes Street Leicester Square 1843, architect – Thomas Little, (RIBA Archives LIT/1/1 + 2).
17 Elisabeth Nagelschmidt (nee Benjamin) in interview with Lynne Walker, in (ed.) *Twentieth Century Architecture*, Vol.2, 'The Modern House Revisited' 1996, 80.
18 Specification for house at Hedgerley Lane, Gerrards Cross, Bucks, 1936, architect – Elisabeth Benjamin, (RIBA Archives, SaG/9/4).
19 Specification for house at Hedgerley Lane (1936).
20 Specification for house at Hedgerley Lane (1936).
21 *Light, Air and Simplicity: House at Gerrards Cross [1936]*, produced and directed by Angela Daniell (London: Momentum, 1997) available at RIBA library, VID0057.
22 Letter from Arnold Colaço-Osorio to Elisabeth Benjamin, 3.10.1936 (RIBA Archives, SaG/9/1).
23 Christopher Powell, *The British Building Industry since 1800: An Economic History* (London: E & FN Spon, 1996), 87–8.

24 'Olympia, 1936', 296.

25 For more on the impact of proprietary specification in the interwar period, and the range of impacts for design, the practice of architecture and society more widely, see Katie Lloyd Thomas, 'This Strange Interloper: building products and the emergence of the architect-shopper in 1930s' Britain' in *Suffragette City: Women, Politics and the Built Environment*, (eds) Elizabeth Darling and Nathaniel R. Walker (London: Routledge, 2019); and 'The Architect as Shopper: Women, electricity, building products and the interwar 'proprietary turn' in the UK' in *Architecture and Feminisms: Economies, ecologies, technologies*, (eds) Hélène Frichot, Catharina Gabrielsson, Helen Runting (London: Routledge, 2018). A monograph on this subject, *The Architect as Shopper: Building Products and the Interwar Proprietary Turn* is in preparation.

26 Immanuel Kant, *The Critique of Pure Reason*, trans. Norman Kemp Smith (London: Macmillan and Co., 1933), 538.

27 Ibid., 539.

28 Aristotle, *The Metaphysics,* trans. Hugh Lawson-Tancred (London: Penguin, 1998), 195.

29 SfB was developed in Sweden in the 1950s and adapted for use in Swedish specifications with the aim of replacing the 'outgrown' trade-based arrangement of the specification with a logical, government-funded system. See L.M. Giertz, *SfB and Its Development 1950–80* (Dublin: CIB/SfB International Bureau, 1982), 5–6.

30 Table 2/3 and the definitions of each of the other tables are taken from the section openers in the *Construction Indexing Manual* and is discussed in Giertz, *SfB and Its Development*, 8–9.

31 In the wonderment of this taxonomy, the thing we apprehend in one great leap, the thing that . . . is demonstrated as the exotic charm of another system of thought, is the limitation of our own, the stark impossibility of thinking that.' Michel Foucault, *The Order of Things* (London: Tavistock Publications, 1970, first publ. in French, 1966), xv.

4 Process

1 Description of the several works to be done in erecting a new chapel and wing buildings in St James Burial Ground, Tottenham Court Road (1791–2), architect – Thomas Hardwick, RIBA Archives AC/HAR/Add/2.

2 Bertrand Gille (ed.), *History of Techniques Vol.2 Part 3, Techniques and Sciences*, trans. J. Brainon, K. Butler et al., (New York, Gordon + Bream Scientific Publishers, 1986), 1146 and 1147.

3 Theophilius, *On Divers Arts*, trans. and introduced by John G. Hawthorne and Cyril Stanley Smith, (New York: Dover Publications, 1979), xxx.

4 Gille, *History of Techniques Vol.2.*, 1149.

5 Graham Harman, *Prince of Networks: Bruno Latour and Metaphysics*, (Melbourne: Re.press, 2009), 6.

6 I am not aware of any remarks upon the proliferation of process-based specification in the 19th century. According to Brian Hanson it is precisely in this period, beginning in the late 18th century, that architects become engaged with the question of the separation between their own 'intellectual' work and the material work of the builders who will realize their forms, in part he suggests because of the cementing of class distinctions between them. According to Hanson, craft emerges as a concern for the first time since the Renaissance for architects such as Pugin and Ruskin, and also awareness of this separation also leads to a marked increase in the degree of control architects attempt to establish over the building process. While he sees the great expansion in the detail and extent of the specification as part of the construction of architects' authority and looks at Bartholomew's 1841 *Specifications for Practical Architecture*, he does not observe the emergence of process-based specifying, which would greatly support his claims. Following his analysis we could add to the economic and industry based causes for this form of clause a more generous interpretation involving an interesting cultural shift towards construction and craft as a valid concern for architectural practice. See Brian Hanson, *Architects and the 'Building World' from Chambers to Ruskin; Constructing Authority* (Cambridge: Cambridge University Press, 2003).

7 Specification for a school house in Hammersmith (1819).

8 The last phrase 'fit for paint' is a performance clause. Specification for the London Fever Hospital, (1833–43), Mocatta, *Specifications*.

9 Specification for a house at Farnham Common (1934).

10 Specification for the Elfrida Rathbone School (1961).

11 Michelmore, Report, 1963.

12 Carter, Report on RIBA proposal for the NBS.

13 Colin McGregor, email correspondence with author, 06.03.2006. McGregor adds that 'there are exceptions to this.'

14 Tony Allot, 'NBS; A Progress Report' in *RIBA Journal*, February 1971, 82.

15 Roland Barthes, 'Plastic' in *Mythologies*, trans. Annette Lavers (London: Vintage, 2000), 97.

16 Muriel Combes, 'Tentative D'ouverture D'une Boite Noire Ce Que Renferme la «Question de la Technique»' in Jean-Marie Vaysse (ed.), *Technique, Monde, Individuation: Heidegger, Simondon, Deleuze* (Hilldesheim: Georg Olms Verlag, 2006), 87.

17 Aristotle, *Metaphysics*, 174.

18 René Descartes, *Discourse on the Method and Meditations*, trans. F.E. Sutcliffe (London: Penguin, 1968), 108–12.

19 Aristotle, *Metaphysics*, 194. Here 'matter' is translated from the Greek term *hyle* which also means wood (and is occasionally used for other materials) and is, as I understand it, a positive term which refers to an instance of the material. Aristotle uses a second term *hypokomenon*, which is translated here as 'substrate' and refers to prime matter. This word literally means 'that which lies beneath' – it has the sense of something behind, that can perhaps be deduced rather than touched.

20 See 'Everything is produced either (i) from a bearer of the same name, as in the case of things produced naturally – an example of this among artefacts is a building, which is produced *from* a building to the extent that it is produced *by* thought, in that the skill is the form of the building,' ibid., 198.

21 Gaston Bachelard, *Water and Dreams: An essay on the imagination of matter*, trans. Edith R. Farrell, (Dallas: Dallas Institute of Humanities and Culture, 1983), 2.

22 According to Ernan McMullin form and matter cannot be separate for Aristotle. It is only with Aquinas that the distinction between them becomes an ontological question. See 'The Concept of Matter' in Ernan McMullin (ed.), *The Concept of Matter in Modern Philosophy* (Notre Dame: University of Notre Dame Press, 1978).

23 See McMullin, 'The Concept of Matter' and Richard Blackwell, 'Descartes' Concept of Matter' in McMullin (ed.) *The Concept of Matter in Modern Philosophy*.

24 For general discussion of the relationships between Descartes' thought and architecture, see Claudia Brodsky Lacour, *Lines of Thought: Discourse, Architectonics, and the Origin of Modern Philosophy* (Durham NC: Duke University Press, 1996).

25 The 'forgetting of operation' is a term Muriel Combes uses in *Simondon. Individu et collectivité*, Paris: PUF, 1999, 17, cited in Toscano, *The Theatre of Production*, 43. 'Veiling' is used by Simondon, and it is not clear to what extent it carries the active sense of concealment and unconcealment which are so important with respect to truth for Heidegger (see the later section of 'The Origin of the Work of Art') and for Nietzsche (see 'On the Uses and Disadvantages of History for Life' in Friedrich Nietzsche, *Untimely Meditations*, trans. R.J. Hollingdale (Cambridge: Cambridge University Press, 1997).

26 Simondon, *L'individu*, 40.

27 Mackenzie, *Transductions*, 48.

28 Simondon, *L'individuation à la lumière des notions de forme et d'information* (Grenoble: Editions Jérôme Millon, 2006), 559. This text is translated into English as 'Allagmatics' in Gilbert Simondon, *Individuation in Light of Notions of Form and Information, Volume II: Supplemental Texts*, trans. Taylor Adkins, (Minneapolis: University of Minnesota Press, 2020), 663–73.

29 Simondon, 'The Genesis of the Individual,' 298.

30 Simondon, *L'individu*, 43.

31 Ibid., 4. As translated in 'The Genesis of the Individual', 300.

32 Simondon, *L'individu*, 40.

33 Ibid., 32, fn.2. As translated in 'The Physico-biological Genesis of the Individual'. (Chapter 1, parts I and II only), trans. Taylor Adkins as and posted at http://fractalontology.wordpress.com /2007/ 10/03/ translation-simondon-and-the-physico-biological-genesis-of-the-individual/ (accessed: 4 December 2008).

34 Theophilius, *On Divers Arts*, 81.

35 Ibid. 83.

36 Ibid. 97.

37 Simondon, *L'individu*, 43–4, as translated in 'The Physico-biological Genesis of the Individual'. Simondon's use here of the residual terms of the hylomorphic schema 'between the matter and the form' should be understood as transitional, as his argument shifts the reader from the hylomorphic model to the allagmatic.

38 According to Ernan McMullin form and matter cannot be separate for Aristotle. It is only with Aquinas that the distinction between them becomes an ontological question. See 'The Concept of Matter' in McMullin (ed.), The *Concept of Matter*, 18.

39 Manuel DeLanda, 'Material Complexity' in Neil Leach, David Turnbull, Chris Williams (eds), *Digital Tectonics* (Chichester: Wiley-Academy, 2004), 21.

40 See Katie Lloyd Thomas, 'Jigging with Concrete: matter and form in the making of Wall One' in Alan Chandler and Remo Pedreschi (eds), *Fabricformwork*, (London: RIBA Publications, 2007), 44–57. And for consideration of these issues in another of Chandler's projects, see Katie Lloyd Thomas, 'Casting Operations and the Description of Process' in *Journal of Architecture*, Vol.20 03 (April 2015) 430–44.

41 Gilles Deleuze and Félix Guattari, *A Thousand Plateaus: Capitalism & Schizophrenia*, trans. Brian Massumi (London: Athlone Press, 1988), 408. Their arguments in this section of the book are made with explicit reference to Simondon's work.

42 Ibid. Note that Deleuze and Guattari use the example of wood to support their discussion of the machinic phylum, rather than Simondon's main example of clay which he demonstrates also has singularities and haecceities.

43 Ibid.

44 Pauline Lefebvre, '"What the Wood Wants to Do": Pragmatist speculations on a response-able architectural practice' in *Architectural Theory Review*, Vol.2.1, 2018, 24–41. Deleuze is cited from a course on Spinoza'e ethics, Gilles Deleuze, 'Gilles Deleuze Spinoza cours du 02/12/80', trans. Christina Rosky, La voix de Gilles Deleuze en ligne (Université Paris 8), http://www2.univ-paris8.fr/deleuze/article.php3?id_article=131 (accessed 5 November 2017).

45 Deleuze and Guattari, *A Thousand Plateaus*, 409.

46 The idea of the artisanal has a powerful legacy for architects. According to Ruskin, writing in *The Stones of Venice*, we can identify two modes of building. Classical building was 'servile' – the slave carried out the orders of the master – while gothic building was 'Christian' (sometimes described as 'democratic'), where the 'system, in confessing the imperfections of the human soul and bestowing "dignity upon the acknowledgement of unworthiness" gave scope for the workman to do as best he could, *following the dictates of his soul*.' Mark Swenarton, *Artisans and Architects: The Ruskinian tradition in architectural thought*, (Basingstoke: Macmillan, 1989), 24, quoting Ruskin. My emphasis. See also John Ruskin, *Stones of Venice*, Jan Morris (ed.), (London: Faber, 1991), 120–3.

47 In explaining this distinction, Simondon gives the example of the 'porosity' of timber which is not in fact a general property of the wood, but a result of timber's implicit forms – the tiny modifications brought about by the pores as they open or close, and of the manner of cutting across or with the grain. See Simondon, *L'individu*, 56.

48 Ibid., 32.

49 Ibid., 52. In a poetic moment Simondon comments in a footnote on this passage that these are expressions of the singularity of the growth of the tree, themselves singularities of another order – the action of the wind or animals – that become information guiding a new operation.

50 Ibid., 59.

51 Deleuze and Guattari, *A Thousand Plateaus*, 410. This seems to be only partially true and depends on the precise metalworking method. Bronze casting is after all Aristotle's favoured example of matter formed.

52 For details of this building see Jeremy Gould, *Modern Houses in Britain 1919–1939* (London: Society of Architectural Historians of Great Britain, 1977) and *Architectural Review*, October 1935, 123–6. The house has recently been restored.

53 The Elfrida Rathbone School for the Educationally Subnormal is in South London and still used, although the boardmarked concrete is now covered with a thick layer of purple paint. It was designed in 1961 and built in 1963/4, by John Bancroft at the LCC, who went on to design the better-known Pimlico School. For more details see John Bancroft, 'Health, Power and Pleasure' in *RIBA Journal*, April 1973, 192–3.

54 According to Eric Classey who worked at the LCC designing schools during this period it was probably put together by a London County Council architect whose sole role was to produce the department's specifications. Interview Nov 2004.

55 Specification for a House at Farnham Common (1934), my emphasis.

56 Ibid.

57 There are two curved walls in the Harding house but they are constructed from blockwork – not concrete – and rendered with a thick plaster.

58 Specification for the Elfrida Rathbone School (1961).

59 Ibid.

60 Ibid., my emphasis.

61 Interview with Alison Smithson, *Zodiac* 4, 1959, 64, cited in Andrew Higgott, *Mediating Modernism: Architectural Cultures in Britain* (London: Routledge, 2007), 92.

62 'On Techno-Aesthetics', Gilbert Simondon, trans. Arne de Boever, in *Parrhesia* 14, www.parrhesiajournal.org/parrhesia14/parrhesia 14_Simondon/pdf, Accessed 18.07.13. Original letter dated 3 July 1982 and published in the original French in the *Papiers du Collège Internationale de Philosophie*, Issue 14. Simondon's interest in architectural examples is explored in more detail in a number of his lectures (accompanied by his wonderful analytical sketches and diagrams) in *L'Invention Dans Les Techniques: Cours et* *conferences*, Gilbert Simondon, (ed.) Jean-Yves Chateau (Paris: Éditions du Seuil, 2005).

63 Elisabeth Shotton, 'Material Imprecision' in Katie Lloyd Thomas (ed.), *Material Matters*, 100.

64 Simondon, *L'individu*, 48.

65 John Protevi, *Political Physics*, 4.

66 Reiser + Umemoto, *Atlas of Novel Tectonics*, (New York: Princeton Architectural Press, 2006), 146.

67 Ibid., 148.

68 In *Du Mode* Simondon does, in a rather different way, suggest that the hylomorphic schema might itself arise out of a particular relationship between worker and work, which he argues is changing with contemporary technological developments and I will discuss this towards the end of this chapter, and in Chapter 7.

69 Simondon, *L'individu*, 49.

70 Simondon, *L'individu*, 59.

71 Discussing Bachelard's *Le Rationalisme Appliqué*, Dominique Lecourt, *Marxism and Epistemology: Bachelard, Canguilhem and Foucault*, trans. Ben Brewster (London: NLB, 1975), 137. Bachelard's epistemological project, and Lecourt's reading of it in terms of dialectical materialism is extremely pertinent to Simondon's philosophy and to this project and would be usefully explored in further research.

72 Ibid., 138.

73 Marian Bowley, *Innovations in Building Materials* (London: Duckworth, 1960), 343–50. Interestingly Bowley explains that the fibreboards industry developed in the UK in the interwar years because plasterers were in short supply and fibreboard could replace their work.

74 For details of the history of MDF production see www.madehow.com/Volume-3/Fiberboard.html (accessed 26 April 2010).

75 Simondon, *L'individu*, 53–4.

76 Smith, *A Search for Structure*, 313.

77 DeLanda, 'Material Complexity', 19.

78 Ibid., 20.

79 Ibid.

80 Marx, *Capital Vol. 1*, 184, my emphasis.

81 Ibid.

82 'Clauses to be introduced into a Specification of Works to be done in Building a Workshop', in Mocatta, *Specifications*, (circa 1833–43).

83 Specification for the Elfrida Rathbone School (1961).

84 Ibid., 31.

85 Ibid.

86 Ibid.

87 Specification for the Elfrida Rathbone School (1961).

88 Brian Massumi, *A User's Guide to Capitalism and Schizophrenia* (Cambridge, Massachusetts: The MIT Press, 1992), 10–20.

5 Performance

1 Ian Chandler, *Building Technology 2; Performance* (London: Mitchell, 1989), 183.

2 George Atkinson, 'Performance Specification,' in *Building*, 17 March 1972, 115–6; 115.

3 Allot, 'NBS; A Progress Report' 82.

4 Stephen Emmitt and David Yeomans, *Specifying Buildings: A Design Management Perspective* (Oxford: Butterworth Heinemann, 2001).

5 Specification of works for a 'Gothic' villa (1879).

6 Specification of works for a smoked sausage manufactory (1934–5).

7 Specification for Air Raid Shelter for Bedales, Hants., architects – Erno Golfinger with Mary Crowley. RIBA Archives, GolER/409/5 (1940–1).

8 Specification for Refurbishment of Scroope Terrace & New Studio Building, University of Cambridge Faculty of Architecture, architects – Mole Architects. Courtesy Mole Architects (2006).

9 Rosen, 'The Future of Specifications Writing – Part II', 206.

10 I am grateful to Astrid Lund, lead of the Technical Content Team at the NBS for sharing her extensive knowledge of the current state of play regarding specifications, telephone interview with author, 8 April 2020.

11 *Performance/Performativity* Bartlett Research Exchange, Bartlett, UCL, London (March 2013).

12 Branko Kolarevic, (ed.), *Performative Architecture: Beyond Instrumentality* (New York: Spon Press, 2005), 3.

13 Ibid.

14 David Leatherbarrow, 'Architecture's Unscripted Performance', in Kolarevic (ed.), *Performative Architecture*, 7.

15 Ibid., 13.

16 Picon looks at performativity in relation to the digital but he is unusual in that he recognizes a broad range of developments, both industrial and practical – from new forms of specification to ecological concerns – that contribute to what can be easily understood as a more general shift. For example: 'The new attention to activity can be traced at various levels. On a technological standpoint, recent evolution points towards the substitution of dynamic performance criteria to static indicators. The ecological character of a building can be apprehended only in dynamic terms. To be green is not a passive attribute; it is the result of a continued action. Similarly, architectural affect is not given once and for all but produced through continuous interaction between subjects and objects.' Antoine Picon, *Digital Culture in Architecture* (Basel: Birkhäuser, 2010), 109.

17 Jon McKenzie, *Perform or Else: From Discipline to Performance* (London: Routledge, 2001), 18.

18 Jean-Francois Lyotard, *The Postmodern Condition: A Report on Knowledge*, trans. Geoff Bennington and Brian Massumi, (Minneapolis: University of Minnesota Press, 1984), 45.

19 The most important and interesting aspect of McKenzie's thesis is that he tracks the possible relationships between a concept of performance which is seen as 'liminal', 'marginal', 'on the edge' and 'a reflexive transgression of social structures' in performance studies but is concerned with efficiency, productivity and normativity in organisational performance management and technological performance, and raises some critical questions for the grip of 'performativity' in many areas of cultural studies.

20 McKenzie, *Perform or Else*, 97.

21 Ibid., 18.

22 Allot, 'NBS: a progress report', 82.

23 NBS: Standard Version (Update 38) (2004), F21, F22.

24 'About Cast Stone' unpaginated webpage http://www.continentalcaststone.com/aboutcaststone.html (accessed 10 June 2009).

25 Carter, 'National Building Specification', 761.

26 Kevan Brassington, 'Pendulum, Ramp and Tortus Slip Resistance Test Methods – Are you confused?' in *NBS Journal* 14, May 2009, 9–10.

27 Ibid., 10.

28 Specification: Project Title: Sainsbury's Maidenhead. For Construction, architect – Chetwoods®. Courtesy Chetwoods®. (2005), H10.

29 Simondon, *Du Mode*, 19, as translated in *On the Mode of Existence of Technical Objects*, 11.

30 Ibid.

31 Bernard Stiegler builds on this notion of technical evolution in *Technics and Time*, Vol.1: *The Fault of Epimetheus*, trans. Richard Beardsworth and George Collins (Stanford: Stanford University Press, 1998).

32 Simondon, *Du Mode*, 15.

33 Simondon, *Du Mode*, 222.

34 Combes 'Tentative D'Ouverture D'une Boite Noire', 75. Combes uses the term 'technique' to mark the shared concerns of Simondon and Heidegger, a move which is more easily made in French than in English – 'The question concerning technology' is translated as 'La Question de la technique' in French, and Simondon's technical objects are 'objects techniques'.

35 Ibid., 218. It is also important for Simondon that it should be possible to make distinctions between living and non-living beings without recourse to their essential vitality.

36 Ibid., 27.

37 Ibid.

38 Ibid., 28.

39 Martin Heidegger, 'The Origin of the Work of Art' in *Poetry, Language, Thought*, trans. Albert Hofstadter (New York: Harper & Row Publishers, 1971), 28.

40 Ibid., my emphasis.

41 Ibid., 64.

42 Martin Heidegger, *The Fundamental Concepts of Metaphysics: World, Finitude, Solitude*, trans. William McNeill and Nicholas Walker (Bloomington: Indiana University Press, 1995), 228.

Here prescription is a direct translation of the noun 'vorschrift' which comes from the verb 'vorschreiben' meaning to prescribe. As in English, the root for this term is to write, but 'schrift' is not an equivalent to 'script' in the sense we use in the theatre. We might also note that in 'The Origin of the Work of Art' a different term is used 'vorzeichnen' for prescribe whose root is 'to draw'. 'vorschreiben' has a more legal sense than 'vorzeichnen' which may explain the difference in Heidegger's terminology, or perhaps the drawing-based term seems more appropriate in an essay about artworks. It may also be that there is no equivalent noun to 'vorschrift' which can be made from 'vorzeichnen'. With thanks to Gerald Niederle for his help with the original terminology.

43 An earlier draft of 'The Origin of the Work of Art' from 1935 suggests there may be more continuity between these two texts. In this version Heidegger also associates hylomorphism with the artwork but does not discuss the question of use. Martin Heidegger, 'The Origin of the Work of Art: First Version', trans. Jerome Veith, in Günther Figal (ed.), *The Heidegger Reader* (Bloomington: Indiana University Press, 2009).

44 Heidegger, *Fundamental Concepts*, 213.

45 Ibid., 228–9.

46 Ibid., 215.

47 Ibid., 219.

48 Ibid.

49 Of course, the blanks in Heidegger's sentence also recall the blanks of the clauses of the specification. It is not the specifics of what the clauses prescribe which is significant but the way in which they do it – the forms of prescription (for use, or through process and so on) that are taken.

50 David Button and Brian Pye (eds), *Glass in Building* (Oxford: Butterworth Architecture, 1993), vii, my emphasis.

51 Michelle Addington and Daniel Schodek (eds), *Smart Materials and Technologies for the Architecture and Design Professions* (Oxford: Architectural Press, 2005), 3.

52 See ibid., 29: 'In the traditional engineering approach the material is understood as an array of physical behaviours. Then in the traditional architectural and general design approach the materials is still conceived as a singular static thing, an artefact.'

53 Bruno Latour, *Pandora's Hope: Essays on the reality of science studies* (Cambridge, Massachusetts: Harvard University Press, 1999), 187.

54 Peter Buchanan, 'When Democracy Builds' in Norman Foster (ed.) *Rebuilding the Reichstag* (London: Weidenfield & Nicholson, 2000), 169.

55 Deborah Ascher Barnstone, *The Transparent State: Architecture and politics in postwar Germany* (London: Routledge, 2005), 125–6.

56 Hisham Elkadi, *Cultures of Glass Architecture* (Aldershot: Ashgate, 2006), 48. He notes it applies to many other buildings including office blocks, which open their ground floors to the gaze but 'exclude people.'

57 Specification of the work, and particulars of the materials to be used in the erection of an asylum for the blind, Bristol (1835), architects – Rickman and Hussey in, a bound collection entitled *Specifications*, RIBA library, EW 2772.

58 Email correspondence with Angela Davidson of Foster and Partners, 14 February 2007.

59 Architectural Specification for Parkside and Blossom Square Kiosks, Potters' Fields Park (2006), architect – DSDHA.

60 John Fernandez, *Material Architecture: Emergent materials for innovative buildings and ecological construction* (London: Architectural Press 2006), 80.

61 In the NBS performance clauses are peppered throughout the work sections. Schumann Smith endeavour to pull different types of clause together and each work section is divided into two type of clause; 'Performance Requirements' and 'Systems, Materials and Fabrication' in which systems and materials are described 'as is'.

62 Specification, The H. . .y Building (2006), prepared by Schumann Smith for . . . Architects. Section H11. A condition of using this specification for this research is that neither building nor architects are identified.

63 Specification for the H. . .y Building (2006), H11.

64 The table was part of a specification in production for a group of office buildings in Coventry (2007) by architects Allies and Morrison, who omitted the values from this table for legal reasons before releasing this information to me.

65 Hans Joachim Gläser, 'The European History of Coatings on Architectural Glazing', http://www.glassfiles.com/library/article.php?id=1051&search=coatings+history&page=1 (accessed 27 July 2009).

66 See Addington and Schodek, *Smart Materials*, 29: 'In the traditional engineering approach the material is understood as an array of physical behaviours. Then in the traditional architectural and general design approach the materials is still conceived as a singular static thing, an artefact.'

67 Smith, *A Search for Structure*, 125.

68 Bernadette Bensaude-Vincent and Isabelle Stengers, *A History of Chemistry*, (Cambridge Massachusetts: Harvard University Press, 1996), 206.

69 Ibid., A History of Chemistry, 205.

70 Bruno Latour, 'Mixing Humans and Nonhumans Together' in William Braham and Jonathan Hale (eds), *Rethinking Technology: A Reader in Architectural Theory* (London: Routledge, 2007), 319.

71 Latour, *Pandora's Hope*, 187.

72 Latour, 'Mixing Humans and Nonhumans Together', 315.

73 Bruno Latour, 'Which politics for which artifacts?' in *Domus*, June 2004, www.bruno-latour.fr/presse/presse_art/GB-06%20DOMUS%2006–04.html (accessed 9 January 2009).

74 See Figure 18.4, ibid., 276. See also the glass manufacturer Romag's much more complex table that includes details of the ammunition and the number of shots that can be withstood. http://www.romag.co.uk/common/attachments/32403%20Romag_BulletRes2ppDS.pdf (accessed 1 August 2013).

75 'Agents can be human or (like the gun) non-human, and each can have goals (or functions, as engineers like to say)' See Latour, *Pandora's Hope*, 180.

76 Ibid. 320.

77 See http://www.romag.co.uk/support-services/architectural-security-support-services/product-testing (accessed 1 August 2013).

78 See Figure 16.3 Test rig, Pilkington Glass Ltd., St Helens, UK in Button and Pye, *Glass in Building*, 258.

79 P.J. Sereda, 'Performance of Building Materials', in *Canadian Building Digest* 115, July 1969. http://irc.nrc-cnrc.gc.ca/pubs/cbd/cbd115_e.html (accessed 10 June 2009).

80 Button and Pye (eds), *Glass in Building*, 121.

81 Sergio de Gaetano, 'Design for Glazing Protection against Terrorist Attacks'. Paper given July 2005 at Glass Perrfomance Days. http://www.glassfiles.com/library/article.php?id=958&search=Gaetano&page=1 (accessed 27 July 2009).

82 Albena Yaneva, *Mapping Controversies in Architecture* (Abingdon: Routledge, 2016), 19.

83 Telephone conversation with Welsh Assembly specification writer, Geoff Taylor of Schumann Smith, May 2008.

84 Latour, 'Mixing Humans and Nonhumans Together', 313.

85 Madeleine Akrich, 'The De-scription of Technical Objects' in Wiebe Bijker and John Law (eds), *Shaping Technology/Building Society: Studies in Sociotechnical change* (Cambridge, Massachusetts: MIT Press, 1992), 222.

86 This well-known distinction, also developed by Puig de la Bellacasa, in her *Matters of Care*, was put forward in Bruno Latour, 'Why Has Critique Run out of Steam? From Matters of Fact to Matters of Concern' in *Critical Inquiry*, Vol.30/2 (2004), 225–48.

87 Latour, *Pandora's Hope*, 190.

6 Systems of Material

1 Miguel de.Beistegui, 'Science and Ontology: From Merleau-Ponty's "reduction" to Simondon's "transduction,"' *Angelaki*, Vol.10, Issue 2, August 2005, 286–87; 118.

2 Simondon, *Du Mode*, 243.
3 Simondon, *L'individu*, 43. As translated in 'The Physico-biological Genesis of the Individual'.
4 Bensaude-Vincent and Stengers, *History of Chemistry*, 206, my emphasis.
5 Andrew Barry, 'Pharmaceutical Matters: The Invention of Informed Materials,' in Mariam Fraser, Sarah Kember, Ceila Lury (eds), *Inventive Life: Approaches to New Vitalism* (London: SAGE, 2006), 59.
6 Ibid., 57.
7 Alfred North Whitehead, *Science and the Modern World* (New York: Mentor Books, 1964), 100, my emphasis.
8 Barry, 'Pharmaceutical Matters', 65.
9 Ibid., 52.
10 Ibid., 58–9.
11 Mackenzie, *Transductions*, 49.
12 Ibid., 288.
13 Ibid., 36. As translated in 'The Physico-biological Genesis of the Individual'.
14 Ibid., 106.
15 Ibid., 107.
16 Simondon also examines the way that individuation of the crystal only occurs at its surface, which he will distinguish from the living being where individuation also has an interior power.
17 Ibid., 107–8.
18 Toscano, *The Theatre of Production*, 142.
19 Ibid., 288.
20 Ibid., fn.1. As translated in 'The Genesis of the Individual', 318, fn.5. Also see fn.13 for a more direct description of photosynthesis as vital individuation.
21 Simondon, *Individuation psychique et collective*, 95, fn.6. As translated in Gilbert Simondon, *Individuation in light of notions of form and information*, trans. Taylor Adkins (Minneapolis: University of Minnesota Press, 2020), fn.7, 394.
22 Ibid., 94, fn.1. As translated in Simondon, *Individuation in light of notions of form and information*, 393, fn.1. The term 'ensemble' or 'whole' has a particular significance in *Du Mode* where it refers to a kind of technical object which comes to prominence in the 20th century. Technical 'elements' are stand-alone technical objects through which man works on nature. Technical 'individuals' are those that seem to replace the work of man, particularly the machines through which early industrialization develops. Technical 'ensembles' are usually groupings of technical elements in which the development of the parts evolves together in specific relation to the ensemble they are part of and involve what Simondon calls an 'associated milieu'. The parts of a technical ensemble form a system, that is to say they are in an informational relation with each other, which has contributed to their specific evolution. While for Simondon it is possible for a man either to control a technical ensemble or become a cog in its functioning, there is a

third possibility, particular to technical ensembles, which is that man can regulate the ensemble, becoming part of a system which is, then constantly inventive. His choice of the term 'ensemble' in this case to mean a set of parts that are not in systemic relation to each other, is confusing, since in *Du Mode* it seems precisely to refer to an individuating technical system.

23 Gilbert Simondon, *L'Invention dans les Techniques: Cours et conferences* (ed.) Jean-Yves Chateau (Paris: Éditions du Seuil, 2005), 278.

24 Ibid., 277.

25 Ibid., 286.

26 Ibid., 287.

27 Frederic Jameson, 'Architecture' in *Postmodernism or, The Cultural Logic of Late Capitalism* (London: Verso, 1991), 113, my emphasis.

28 For emphasis on the aesthetic experience of architecture see for example Juhani Pallasmaa who argues that with the 'hegemony of vision' contemporary architecture has neglected the body, and pays particular attention to materiality – suggesting for example that building with mud and clay engages with 'the tacit wisdom of the body' or that 'a weakened sense of materiality' may be another outcome of a scopic regime. Juhani Pallasmaa, *The Eyes of the Skin: Architecture and the Senses* (Chichester: Wiley – Academy, 2005), 17, 26, 31, and also Steven Holl, Juhani Pallasmaa, Alberto Péréz-Gómez, *Questions of Perception: Phenomenology of Architecture* (Tokyo: a+u Publishing, 2006).

29 Nicholas Coetzer, 'Between Birds' Nests and Manor Houses: Edwardian Cape Town and the Political Nature of Building Materials' in Katie Lloyd Thomas (ed.), *Material Matters: Architecture and Material Practice* (London: Routledge, 2007), 191.

30 See Andrew Benjamin, 'Notes on the Surfacing of Walls: Nox, Kiesler, Semper,' in Lars Spuybroek, *Nox: Machining Architecture* (London: Thames and Hudson, 2004) and 'Plans to Matter' in Katie Lloyd Thomas (ed.), *Material Matters: Architecture and Material Practice*, (London: Routledge, 2007).

31 Ákos Moravánsky, *Metamorphism: Material change in architecture* (Basel: Birkhäuser, 2018).

32 Colin Rowe and Robert Slutzky, 'Transparency: Literal and Phenomenal' in *Perspecta*, Vol.8, 1963, 53.

33 Eisenman explains that he uses white card for the design of House I because it is 'closer to an abstract plane than say a natural wood or a cut stone wall, and therefore reduces existing meaning'. Peter Eisenman, *Eisenman Inside Out: Selected Writings 1963–1988* (Newhaven: Yale University Press, 2004), 30.

34 Ibid., 89.

35 Mark Jarzombek, 'The Quadrivium Industrial Complex', *e-flux architecture*, 'Overgrowth' www.e-flux.com/architecture/overgrowth/296508/the-quadrivium-industrial-complex/ (accessed 3 May 2021).

36 Reyner Banham, *A Concrete Atlantis: US industrial building and European modern architecture 1900–1925*. (Cambridge, Massachusetts: The MIT Press, 1986).

37 See in particular, Sérgio Ferro, 'Concrete as Weapon,' trans. Alice Fiuza and Silke Kapp, *Harvard Design Magazine* No.46, F/W 2018 (insert), and also, Sérgio Ferro, *Dessin/Chantier: an introduction*, trans. Ricardo Agarez and Silke Kapp in (eds) Katie Lloyd Thomas, Tilo Amhoff and Nick Beech (London: Routledge, 2016), and for a critique of contemporary star architects' design informed by Ferro's work, see Pedro Fiori Arantes, *The Rent of Form: Architecture and Labor in the Digital Age*, trans. Adrianna Kauffmann (Minneapolis: University of Minnesota Press, 2019).

38 Matthew Gandy, *Concrete and Clay: Reworking nature in New York City* (Cambridge, Massachusetts: The MIT Press, 2003).

39 Jane Hutton, *Reciprocal Landscapes: Stories of material movements* (London: Routledge, 2019). See also her guest edited issue, Jane Hutton (ed.), *Material Culture: Assembling and disassembling landscapes*, Landscript 5 (Berlin: Jovis, 2017).

40 For these papers see *7ICCH Proceedings* (CRC Balkema, forthcoming).

41 Hannah le Roux and Gabrielle Hecht, 'Bad Earth', www.e-flux.com/architecture/accumulation/345106/bad-earth/ (accessed 3 May 2021). See also Kim Förster's research on the global cement industry 'Cement: Building material of the anthropocene, www.kimfoerster.com/index.php?/current-project/young–giroux (accessed 3 May 2021).

42 Ibid., 24, my emphasis.

43 Bowyer, *Practical Specification Writing*, 55, my emphasis.

44 Specification for the H. . .y Building (2006).

45 Ibid.

46 Specification for the Q. . .t Building, prepared by Schumann Smith for . . . Architects. Courtesy Schumann Smith. (2005).

47 Specification for the H. . .y Building, (2006).

48 Ruben Suare, 'Innovation Through Accountability in the Design and Manufacturing of Material Effects' in. Branko Kolarevic and Kevin Klinger (eds), *Manufacturing Material Effects: Rethinking Design and Making in Architecture*, (London: Routledge, 2008), 57.

49 Ibid.

50 Elizabeth Diller presenting the mock up, citied in Suare, 'Innovation Through Accountability', 58.

51 Specification of MATERIALS and WORKMANSHIP For Greenhouse and Summerhouse at Richmond Terrace Gardens (2005), architects – Caroe & Partners, specification prepared by René Tobe.

52 Architectural Specification for Parkside and Blossom Square Kiosks, Potters' Fields Park (2006), architect – DSDHA. Courtesy DSDHA.

53 Ibid.

54 Ibid.

55 Interview with Simon Tucker of Coterell Vermeulen, 2 April 2007.

56 Schedule attached to letter from London Borough of Islington Control, in respect of 9 Stock Orchard Street, 1 July 1998. My emphasis.

57 Jeremy Till and Sarah Wigglesworth, 'The Future is Hairy', in Jonathan Hill (ed.), *Architecture: The Subject is Matter* (London: Routledge, 2001), 16.

58 'Extrafunctional' is a term used by Elizabeth Grosz in her essay 'Architectures of Excess' in *Architecture from the Outside* (Cambridge, Massachusetts: MIT Press, 2001), 163, first published in (ed.) Cynthia Davidson, *Anymore* (Cambridge MA: MIT Press, 2000). A more detailed account of the notion of excess in relation to the straw bale house is given in my 'The excessive materiality of Stock Orchard Street: towards a feminist material practice' in Sarah Wigglesworth (ed.), *Around and About Stock Orchard Street* (London: Routledge, 2010), 118–31.

59 This was also the case with the upholstered wall, an innovative, untested material which similarly performed an insulating role which could not be recognized in the terms of building control. Thus, its functional 'excess' is produced by statutory frameworks, as opposed to its literal or physical role as part of the building's functioning structure.

60 Notes of Meeting between Seyan/Andrews from London Borough of Islington Building Control and Sarah Wigglesworth and Jeremy Till, 24 June 1998.

7 Going Into the Mould

1 Gilbert Simondon, On the Mode of Existence of Technical Objects, trans. Cécile Malaspina and John Rogove (Minneapolis: Univocal, 2017), 252.

2 These arguments are also developed in remarkable text 'Individuation and Invention' recently made available in English, and published with other supplemental texts to accompany *Individuation in Light of Notions of Form and Information*; Gilbert Simondon, *Individuation in Light of Notions of Form and Information, Volume II Supplemental Texts*, trans. Taylor Adkins (Minneapolis: University of Minnesota Press, 2020).

3 Simondon, On the Mode of Existence of Technical Objects, 119.

4 Ibid. 249.

5 Forty, *Concrete and Culture*, 44.

6 Specification for the Elfrida Rathbone School (1961).

7 *National Building Specification: Standard Version (Update 38)* (London: RIBA Enterprises Ltd. 2004).

8 Adam Caruso, 'Towards an Ontology of Construction' in Caruso St John Architects, *Knitting Weaving Wrapping Pressing* (Luzern: Edition Architekturgalerie Luzern, 2002), 7.

9 'Simondon, 'The Genesis of the Individual', 312.

10 Ibid.

11 Simondon, 'Individuation and Invention' in *Individuation in Light of Notions of Form and Information; Volume II Supplemental Texts*, trans. Taylor Adkins (Minneapolis: University of Minnesota Press, 2020), 412–34.

12 Simondon, 'Individuation and Invention', 419.

13 Ibid., 415.
14 Simondon, *On the Mode of Existence of Technical Objects*, 30.
15 Ibid., 183.
16 Ibid., 49.
17 Ibid., 40.
18 Georges Canguilhem, 'Machine and Organism' in J. Crary and S. Kwinter (eds) *Incorporations*, New York: Zone Books, 1992, first published in French, 1947, 63–4. As cited in Daniela Voss 'Invention and Capture: a critique of Simondon' in *Culture, Theory and Critique* Vol. 60, 3–4, 279–99 (279).
19 Muriel Combes, *Gilbert Simondon and the Philosophy of the Transindividual*, trans. Thomas LaMarre (Cambridge, Massachusetts: The MIT Press, 2013), 67.
20 Simondon, *On the Mode of Existence of Technical Objects*, 43.
21 Ibid., 250–1.
22 Ibid., 17.
23 Ibid., 251.
24 Paolo Virno, in 'Reading Gilbert Simondon: Transindivuality, technical activity and reification' an interview with Jun Fujita Hirose, trans. Matteo Mandarani, *Radical Philosophy*, No. 136, March/April 2006, 36.
25 Simondon, *On the Mode of Existence of Technical Objects*, 50.
26 Ibid., 251.
27 Ibid., 258.
28 Ibid., 250.
29 Ibid., 247.
30 For example, see, Simondon, 'Individuation and Invention', 413–14.
31 Simondon, *On the Mode of Existence of Technical Objects*, 247.
32 Ibid.
33 Ibid., 248.
34 Ibid.
35 Ibid., 254–5.
36 Ibid., 249.
37 Ibid., 248–9.
38 Ibid., 249.
39 Ibid., 256.
40 Simondon, 'Individuation and Invention', 432–3.
41 In this view Daniela Voss follows Alberto Toscano, using his terms to provide the title of her essay, Daniela Voss 'Invention and Capture: a critique of Simondon' in *Culture, Theory and Critique* Vol. 60, 3–4, 279–99 (279).
42 Combes, *Gilbert Simondon and the Philosophy of the Transindividual*, 76.
43 Simondon, *On the Mode of Existence of Technical Objects*, 254.
44 Ibid., 257.

45 Ibid., 256.

46 Alberto Toscano, 'Technical Culture and the Limits of Interaction: A Note on Simondon' in (eds) Joke Brouwer and Arjen Mulder, *Interact of Die!* (Rotterdam: V2_Publishing/NAi Publishers, 2007), 205.

47 Simondon, *Du Mode*, as cited in Toscano, ibid. in *Simondon, On the Mode of Existence of Technical Objects*, 255.

48 This point is also made by Léopold Lambert and described as a kind of 'violence' in his short article 'For an Allagmatic Architecture: Introduction to the work of Gilbert Simondon' available at https://thefunambulist.net/architecture/simondon-episode-01-for-an-allagmatic-architecture-introduction-to-the-work-of-gilbert-simondon (accessed 25 April 2021).

49 For more on this film see (eds) Antje Ehmann and Kodwo Eshun, *Harun Farocki: Against What? Against Whom?* (London: Koenig Books, 2009) and Katie Lloyd Thomas and Nick Beech, 'Into the Hidden Abode: Architecture, Production, Process' in *Architecture and Culture* Vol. 3/3 (November 2015), 271–9.

50 For Sérgio Ferro see for example, *Concrete as Weapon*, trans. Alice Fiuza and Silke Kapp, *Harvard Design Magazine* No. 46, F/W 2018 (insert); *Dessin/Chantier: an introduction*, trans. Ricardo Agarez and Silke Kapp in (eds) Katie Lloyd Thomas, Tilo Amhoff and Nick Beech, *Industries of Architecture* (London: Routledge, 2016), for Christine Wall, *An Architecture of Parts: Architects, Building Workers and Industrialisation in Britain 1940–1970* (London: Routledge, 2013), for Pedro Fiori Arantes, *The Rent of Form: Architecture and Labor in the Digital Age*, trans. Adrianna Kauffmann (Minneapolis: University of Minnesota Press, 2019) and for Linda Clarke, *Building Capitalism: Historical Change and the Labour Process in the Production of the Built Environment* (London: Routledge, 1992).

51 Voss, 'Invention and Capture: a critique of Simondon', 297.

52 Combes, *Gilbert Simondon and the Philosophy of the Transindividual*, 77.

53 For more on Wall One, see Alan Chandler, 'A Philosophy of Engagement' in (ed.) Katie Lloyd Thomas, *Material Matters: Architecture and Material Practice* (London: Routledge, 2007) 115–24, and Katie Lloyd Thomas, 'Jigging with Concrete: matter and form in the making of Wall One' in Alan Chandler and Remo Pedreschi (eds), *Fabricformwork*, (London: RIBA Publications, 2007) 44–57.

54 For more on HBBE see http://bbe.ac.uk (accessed 22 April 2020).

55 Tom Roberts, 'Thinking technology for the Anthropocene: encountering 3D printing through the philosophy of Gilbert Simondon' in *Cultural Geographies*, Vol. 24.4 (2017), 539–54; 541, 547.

56 See Helena Westerlind, 'Concrete Deposition: Choreographing Flow', in (eds) Bob Sheil, Mette Ramsgaard Thomsen, Martin Tamke and Sean Hanna, *Design Transactions: Rethinking information modeling for a new material age*, (London: UCL Press, 2020), 196–7.

57 Flavio Bevilacqua, *Before Design Theory: Conceptual basis for the development of a theory of design and digital manufacturing based on the relationships between design and Simondon's individuation theory* (self-published: Amazon UK, 2019).

58 Stefano Mazzilli-Daeschel, 'Simondon and the Maker Movement' in *Culture, Theory and Critique*, 60:3–4 (2019), 237–49. He cites Andrea Bardin, *Epistemology and Political Philosophy in Gilbert Simondon: Individuation, Technics, Social Systems* (Dordrecht: Springer, 2015).

59 Voss, 'Invention and Capture: a critique of Simondon', 288. Voss pulls out the wonderful term 'cumulative participation' from Gilbert Simondon, *Imagination et invention*, (ed.) Nathalie Simondon (Chatou: Les Éditions de La Transparence), 163.

60 Thomas LaMarre, 'Afterword: Humans and Machines', in Combes, *Gilbert Simondon and the Philosophy of the Transindividual*, 103.

61 João Marcos de Almeida Lopes, 'Architecture as ensemble: A matter of method', trans. Alice Fiuza, in (eds) Katie Lloyd Thomas, Nick Beech and Tilo Amhoff, *Industries of Architecture* (London Routledge, 2016), 106–13; 110.

62 Simondon, On the Mode of Existence of Technical Objects, 109.

63 Lopes, 'Architecture as ensemble: A matter of method', 109.

64 Ibid.

65 As cited in ibid.,107. Karl Marx, *Capital: A Critique of Political Economy, Vol.I – The Process of Capitalist Production*, trans. Samuel Moore and Edward Aveling, (ed.) Ernest Untermann (Chicago: Charles H Kerr and Company, 1906), 406, fn.2 (Chapter 15).

66 Exhibition leaflet, USINA_ CTAH, curated by Davide Sacconi, *Second Edition of the Biennale d'Architecture d'Orléans*, Collégiale Saint-PierreLe-Puellier, Orléans, 2019. For more on the work of USINA_ CTAH, see http://www.usina-ctah.org.br (accessed 22 April 2020); *Entre-Espaço*, director Cristina Beskow (2016); Pedro Fiori Arantes, 'Reinventing the Building Site' in (eds) Elisabetta Andreoli and Adrian Forty, *Brazil's Modern Architecture* (London and New York: Phaidon, 2004), 193–201.

67 Lopes, 'Architecture as ensemble: A matter of method', 111.

68 For these projects see www.bureaus.dk/specifications and www.bureaus.dk/2020/05/20/impossible-instructions (accessed 5 June 2020).

69 Brady Burroughs, Architectural Flirtations: A Love Storey (Stockholm: Arkitektur-och designcentrum, 2016), 296–7.

70 See https://www.thenbs.com (accessed 22 April 2020).

71 There is today a growing corpus of work that pays attention to the effects of changing techniques and technologies, regulation and financial instruments on architecture. I am thinking for example, of the work of Peggy Deamer, such as (ed.) *The Architect as Worker: Immaterial Labor, the Creative Class and the Politics of Design* (London: Bloomsbury, 2015) and of the members of the architectural history collective Aggregate, such as (eds) Zeynep Çelik Alexander and John May *Design Technics: Archaeologies of Architectural Practice* (Minneapolis: University of Minnesota Press, 2020). However (as is a limitation of this book) this work can risk emphasizing architecture's constitutive seam at the expense of architectural design.

72 Some practitioners engaging inventively with these issues have included Kennedy Violich Architects particularly in their project with the San Francisco

Plasterers' Union. See Sheila Kennedy and Christoph Gruneburg, *Material Misuse: Kennedy and Violich Architecture* (London: Architectural Association Publications, 2001). See also Muf's Stoke project that worked with fabricators from local sanitaryware and porcelain crockery manufacturers in (eds) Katherine Shonfield, Rosa Ainley and Adrian Dannatt *This Is What We Do: A Muf Manual* (London: Batsford, 2001) or Katherine Shonfield's marvellous analysis of the cavity wall and her related installations in 'Why does this flat leak?' in Katherine Shonfield, *Walls Have Feelings: Architecture, Film and the City* (London: Routledge, 2000). Some architects and designers play with the possibilities of planning law, as by David Knight, Finn Williams and Ulf Hackauf in 'Building without Bureaucracy', in *Architecture d'Aujourd'hui* No. 378, June-July 2010, while Liam Ross explores building regulations in his inventive design research, see for example, Liam Ross, 'On contradictory regulations' in *Architectural Research Quarterly* 16/3, 2012, 205–9 and 'Regulatory spaces, physical and metaphorical: on the legal and spatial occupation of fire-safety regulation' in (eds) Lloyd Thomas, Amhoff and Beech, *Industries of Architecture*, 235–44. As Rob Imrie and Emma Street have argued in *Architectural Design and Regulation* (Oxford: Wiley-Blackwell, 2011) architects can engage with regulatory practices rather than consider them as simply imposed.

Bibliography

Books and Articles

Addington, Michelle and Daniel Schodek (eds). *Smart Materials and Technologies for the Architecture and Design Professions*, Oxford: Architectural Press, 2005.

Agarez, Ricardo. *Algarve Building: Modernism, Regionalism and Architecture in the South of Portugal, 1925–1965,* London: Routledge, 2016.

Akrich, Madeleine. 'The De-scription of Technical Objects' in Wiebe Bijker and John Law (eds), *Shaping Technology/Building Society: Studies in Sociotechnical change*, Cambridge, Massachusetts: MIT Press, 1992.

Alexander, Zeynep Çelik. 'Scanning: a technical history of form', in Zeynep Çelik Alexander and John May (eds), *Design Technics: Archaeologies of architectural practice*, Minneapolis: University of Minnesota Press, 2020.

Alexander, Zeynep Çelik and John May (eds). *Design Technics: Archaeologies of Architectural Practice*, Minneapolis: University of Minnesota Press, 2020.

Allot, Tony. 'NBS; A Progress Report' in *RIBA Journal*, February 1971, 82.

Amhoff, Tilo. *Adapting the architect's products to the capitalist building production: The development of the Legal Obligations, Building Specifications and Working Drawings in the first half of the nineteenth-century in England*, unpublished dissertation submitted for the MSc in Architectural History, The Bartlett, University College London, 2003/2004.

Ankele, Gudrun and Danielle Zyman (eds). *Los Carpinteros: Handwork, Constructing the World*, Köln: Walter König, 2011.

Antonelli, Paola with Anna Burckhardt (eds). *The Neri Oxman Material Ecology Catalogue*, New York: The Museum of Modern Art, 2020.

Arantes, Pedro Fiori. *The Rent of Form: Architecture and Labor in the Digital Age*, trans. Adrianna Kauffmann, Minneapolis: University of Minnesota Press, 2019.

Arantes, Pedro Fiori. 'Reinventing the Building Site' in Elisabetta Andreoli and Adrian Forty (eds), *Brazil's Modern Architecture,* London and New York: Phaidon, 2004; 193–201.

Aristotle. *The Metaphysics*, trans. Hugh Lawson-Tancred, London: Penguin, 1998.

Armstrong, Rachel. *Vibrant Architecture: Matter as a Codesigner of Living Structures*, Warsaw: De Gruyter Open, 2015.

Ascher Barnstone, Deborah. *The Transparent State: Architecture and politics in postwar Germany*, London: Routledge, 2005.

Atkinson, George. 'Performance Specification,' *Building*, 17 March 1972; 115–6.

Bachelard, Gaston. *Water and Dreams: An essay on the imagination of matter*, trans. Edith R. Farrell, Dallas: Dallas Institute of Humanities and Culture, 1983.

Ball, Michael. *Rebuilding Construction,* London: Routledge, 1988.
Ballard-Bell, Victoria with Patrick Rand. *Materials for Architectural Design,* London: Laurence King Publishing, 2006.
Bancroft, John. 'Health, Power and Pleasure' in *RIBA Journal,* April 1973, 192–3.
Banham, Reyner. *A Concrete Atlantis: US industrial building and European modern architecture 1900–1925,* Cambridge, Massachusetts: The MIT Press, 1986.
Bardin, Andrea. *Epistemology and Political Philosophy in Gilbert Simondon: Individuation, Technics, Social Systems,* Dordrecht: Springer, 2015.
Barry, Andrew. 'Pharmaceutical Matters: The Invention of Informed Materials,' in Mariam Fraser, Sarah Kember, Ceila Lury (eds), *Inventive Life: Approaches to New Vitalism,* London: SAGE, 2006.
Barthélémy, Jean-Hughes. 'Encyclopedism' in 'Fifty Key Terms in the Works of Gilbert Simondon', trans. Arne de Boever in Arne de Boever, Alex Murray, Jon Rogge and Ashley Woodward (eds), *Gilbert Simondon: Being and Technology,* Edinburgh: Edinburgh University Press, 2012.
Barthes, Roland. 'Plastic' in *Mythologies,* trans. Annette Lavers, London: Vintage, 2000.
Bartholomew, Alfred. *Specifications for Practical Architecture,* London: John Williams & Co., 1840.
Beistegui, Miguel de. *Truth and Genesis,* Bloomington: Indiana University Press, 2004.
Beistegui, Miguel de. 'Science and Ontology: From Merleau-Ponty's 'Reduction' to Simondon's 'Transduction', *Angelaki,* 10 (2), August 2005; 286–287.
Bennett, Jane. *Vibrant Matter: A political ecology of things,* Durham, N.C.: Duke Press, 2010.
Benjamin, Andrew, 'Notes on the Surfacing of Walls: Nox, Kiesler, Semper,' in Lars Spuybroek, *Nox: Machining Architecture,* London: Thames and Hudson, 2004.
Benjamin, Andrew. 'Plans to Matter' in Katie Lloyd Thomas (ed.), *Material Matters: Architecture and Material Practice,* London: Routledge, 2007.
Benjamin, Elisabeth. 'Elisabeth Nagelschmidt (nee Benjamin) in interview with Lynne Walker', in *Twentieth Century Archtiecture* 2, 'The Modern House Revisited' 1996.
Bensaude-Vincent, Bernadette and Isabelle Stengers. *A History of Chemistry,* Cambridge Massachusetts: Harvard University Press, 1996.
Beskow, Cristina. Director, *Entre-Espaço,* 2016.
Bevilacqua, Flavio. *Before Design Theory: Conceptual basis for the development of a theory of design and digital manufacturing based on the relationships between design and Simondon's individuation theory,* self-published: Amazon UK, 2019.
Blackwell, Richard. 'Descartes' Concept of Matter' in Ernan McMullin (ed.), *The Concept of Matter in Modern Philosophy,* Notre Dame: University of Notre Dame Press, 1978.
Bowley, Marian. *Innovations in Building Materials,* London: Duckworth, 1960.
Bowyer, Jack. *A History of Building,* Eastbourne: Orion Books, 1983.
Brassington, Kevan. 'Pendulum, Ramp and Tortus Slip Resistance Test Methods – Are you confused?' in *NBS Journal* 14, May 2009.
Brodsky Lacour, Claudia. *Lines of Thought: Discourse, Architectonics, and the Origin of Modern Philosophy,* Durham NC: Duke University Press, 1996.

Buchanan, Peter. 'When Democracy Builds' in Norman Foster (ed.), *Rebuilding the Reichstag*, London: Weidenfield & Nicholson, 2000.
Burroughs, Brady. *Architectural Flirtations: A Love Storey*, Stockholm: Arkitektur- och designcentrum, 2016.
Button, David and Brian Pye (eds). *Glass in Building*, Oxford: Butterworth Architecture, 1993.
Canguilhem, Georges. 'Machine and Organism' in J. Crary and S. Kwinter (eds), *Incorporations*, New York: Zone Books, 1992, first published in French, 1947; 63–64.
Carter, John. Report on RIBA proposal for the NBS, November 1968, RIBA Library, RIBA Services Working Group Papers, Box 1.
Carter, John. 'National Building Specification', *The Architects' Journal Information Library* 19 March 1969; 759.
Caruso, Adam. 'Towards an Ontology of Construction' in Caruso St John Architects, *Knitting Weaving Wrapping Pressing*, Luzern: Edition Architekturgalerie Luzern, 2002.
Chandler, Alan. 'A Philosophy of Engagement' in Katie Lloyd Thomas (ed.), *Material Matters: Architecture and Material Practice,* London: Routledge, 2007; 115–124.
Chandler, Ian. *Building Technology 2; Performance,* London: Mitchell, 1989.
Chew, Sing C. *Logs for Capital: The Timber Industry and Capitalist Enterprise in the 19th Century*, Westport: Greenwood Press, 1992.
Clarke, Linda. *Building Capitalism: Historical Change and the Labour Process in the Production of the Built Environment*, London: Routledge, 1992.
Colaço-Osorio, Arnold. Letter to Elisabeth Benjamin, 3 October 1936 (RIBA Archives, SaG/9/1).
Combes, Muriel. *Simondon. Individu et collectivité*, Paris: PUF, 1999.
Combes, Muriel. *Gilbert Simondon and the Philosophy of the Transindividual*, trans. Thomas LaMarre, Cambridge, Massachusetts: The MIT Press, 2013.
Combes, Muriel. 'Tentative D'ouverture D'une Boite Noire Ce Que Renferme la «Question de la Technique»' in Jean-Marie Vaysse (ed.), *Technique, Monde, Individuation: Heidegger, Simondon, Deleuze*, Hilldesheim: Georg Olms Verlag, 2006.
Construction Indexing Manual, London: RIBA Publications 1969.
Davis, Howard. *The Culture of Building*, Oxford: Oxford University Press, 2006.
Deamer, Peggy (ed.). *The Architect as Worker: Immaterial Labor, the Creative Class and the Politics of Design,* London: Bloomsbury, 2015.
De Gaetano, Sergio. 'Design for Glazing Protection against Terrorist Attacks.' Paper given July 2005 at Glass Perrfomance Days. http://www.glassfiles.com/library/article.php?id=958&search=Gaetano&page=1 (accessed 27 July 2009).
DeLanda Manuel. 'Material Complexity' in Neil Leach, David Turnbull, Chris Williams (eds), *Digital Tectonics*, Chichester: Wiley-Academy, 2004.
Deleuze, Gilles and Félix Guattari. *A Thousand Plateaus: Capitalism & Schizophrenia*, trans. Brian Massumi, London: Athlone Press, 1988.
Descartes, René. *Discourse on the Method and Meditations*, trans. F.E. Sutcliffe, London: Penguin, 1968.
Donaldson, Thomas Leverton. *Handbook of Specifications: or, practical guide to the Architect, Engineer, Surveyor, and Builder, in drawing up specifications and contracts for works and constructions*, London: Atchley and Co., 1859.

Donaldson, Thomas Leverton. *Handbook of Specifications*, London: Lockwood and Co., 1860.
Ehmann, Antje and Kodwo Eshun (eds). *Harun Farocki: Against What? Against Whom?* (London: Koenig Books, 2009).
Eisenman, Peter. *Eisenman Inside Out: Selected Writings 1963–1988*, Newhaven: Yale University Press, 2004.
Elkadi, Hisham. *Cultures of Glass Architecture*, Aldershot: Ashgate, 2006.
Emmitt, Stephen and David Yeomans, *Specifying Buildings: A Design Management Perspective* Oxford: Butterworth Heinemann, 2001.
Fernandez, John. *Material Architecture: Emergent materials for innovative buildings and ecological construction*, London: Architectural Press 2006.
Ferro, Sérgio. *O canteiro e o desenho*, São Paulo: Projeto Editores, 1979.
Ferro, Sérgio. *Dessin/chantier*, Grenoble: Éditions de la Villette, 2005.
Ferro, Sérgio. *Concrete as Weapon*, trans. Alice Fiuza and Silke Kapp, *Harvard Design Magazine No.46*, F/W 2018 (insert).
Ferro, Sérgio. 'Dessin/Chantier: an introduction', trans. Ricardo Agarez and Silke Kapp, in Katie Lloyd Thomas, Tilo Amhoff and Nick Beech (eds), *Industries of Architecture*, London: Routledge, 2016.
Foucault, Michel. *The Order of Things* (1966), Tavistock Publications, London, 1974.
Foucault, Michel. *The Archaeology of Knowledge* trans. A. M. Sheridan Smith, London: Routledge, 1972.
Frichot, Hélène. *Creative Ecologies: Theorizing the Practice of Architecture*, London: Bloomsbury, 2019.
Gandy, Matthew. *Concrete and Clay: Reworking nature in New York City*, Cambridge, Massachusetts: The MIT Press, 2003.
Garelli, Jacques. *Gilbert Simondon*, Paris: Albin Michel, 1994.
Gelder, John. *Specifying Architecture: A Guide to Professional Practice*, Milsons Point: Construction Information Systems Australia, 2001.
Mr George report, Panel meeting notes, 1963, RIBA Library, Specifications Panel Papers, Box 1.
Giedion, Sigfried. *Mechanization Takes Command: A contribution to anonymous history*, Minneapolis: University of Minnesota Press, 2013; first published 1948.
Giertz, L.M. *SfB and Its Development 1950–80*, Dublin: CIB/SfB International Bureau, 1982.
Gille, Bertrand (ed.). *History of Techniques Vol.2 Part 3, Techniques and Sciences*, trans. J. Brainon, K. Butler et al., New York, Gordon + Bream Scientific Publishers, 1986.
Gläser, Hans Joachim. 'The European History of Coatings on Architectural Glazing,' http://www.glassfiles.com/library/article.php?id=1051&search=coatings+history&page=1 (accessed 27 July 2009).
Golding, Maurice in notes to Anthony Laing, chair of the Specifications Panel, 9 July 1964, RIBA Library, Specifications Panel Papers, Box 1.
Gould, Jeremy. *Modern Houses in Britain 1919–1939*, London: Society of Architectural Historians of Great Britain, 1977.
Grosz, Elizabeth. *Architecture from the Outside*, Cambridge, Massachusetts: MIT Press, 2001.
Hagan, Susannah. *Taking Shape: A new contract between architecture and nature*, Oxford: Architectural Press, 2001.

Hanson, Brian. *Architects and the 'Building World' from Chambers to Ruskin; Constructing Authority*, Cambridge: Cambridge University Press, 2003.
Harman, Graham, *Prince of Networks: Bruno Latour and Metaphysics*, Melbourne: Re.press, 2009.
Hayward, Mark and Bernard Dionysius Geoghegan. 'Introduction: Catching Up With Simondon', in *SubStance* 41, No 3, 2012 (Issue 129); 3–15.
Heidegger, Martin. *The Fundamental Concepts of Metaphysics: World, Finitude, Solitude*, trans. William McNeill and Nicholas Walker, Bloomington: Indiana University Press, 1995.
Heidegger, Martin. 'The Origin of the Work of Art' in *Poetry, Language, Thought*, trans. Albert Hofstadter New York: Harper & Row Publishers, 1971.
Heidegger, Martin. 'The Origin of the Work of Art: First Version', trans. Jerome Veith, in Günther Figal (ed.), *The Heidegger Reader*, Bloomington: Indiana University Press, 2009.
Higgott, Andrew. *Mediating Modernism: Architectural Cultures in Britain*, London: Routledge, 2007.
Hill, Jonathan (ed.). *Architecture:* The subject is Matter, London: Routledge, 2001.
The Holy Bible, Revised Standard Version, New York: Collins, 1971.
Holl, Steven, J. Pallasmaa, A. Péréz-Gómez. *Questions of Perception: Phenomenology of Architecture*, Tokyo: a+u Publishing, 2006.
Hurx, Merlijn. *Architecture as profession: The origins of architectural practice in the Low Countries in the fifteenth century*, Brussels: Brepols Publishers, 2017.
Hutton, Jane (ed.). *Material Culture: Assembling and disassembling landscapes, Landscript 5*, Berlin: Jovis, 2017.
Hutton, Jane. *Reciprocal Landscapes: Stories of material movements*, London: Routledge, 2019.
Imrie, Rob and Emma Street, *Architectural Design and Regulation*, Oxford: Wiley-Blackwell, 2011.
Jaeggi, Annemarie. *Fagus: Industrial Culture from Werkbund to Bauhaus*, New York: Princeton Architectural Press, 2014.
Jameson, Frederic. *Postmodernism or, The Cultural Logic of Late Capitalism*, London: Verso, 1991.
Jarzombek, Mark. 'The Quadrivium Industrial Complex,' *e-flux architecture*, 'Overgrowth' www.e-flux.com/architecture/overgrowth/296508/the-quadrivium-industrial-complex/ (accessed 3 May 2021).
Kant, Immanuel. *The Critique of Pure Reason*, trans. Norman Kemp Smith, London: Macmillan and Co., 1933.
Kapp, Silke, Katie Lloyd Thomas and João Marcos de Almeida Lopes. 'How to Look at Architecture from Below', an introduction to Sérgio Ferro, 'Concrete as Weapon', in *Harvard Design Guide*, No.46 – F/W 2018.
Kennedy, Sheila and Christoph Gruneburg. *Material Misuse: Kennedy and Violich Architecture*, London: Architectural Association Publications, 2001.
Knight, David, Finn Williams and Ulf Hackauf in 'Building without Bureaucracy', in *'Architecture d'Aujourd'hui* No.378, June–July 2010.
Knightly Brown, Andrew. 'Stone (Extracting Blood From)' from *The Spec Man Blog*, 13 August 2012, www.andrewkb.net/blog/20120813.htm2013 (accessed 2 November 2013).
Kolarevic, Branko (ed.). *Performative Architecture: Beyond Instrumentality*, New York: Spon Press, 2005.

Laing, Anthony. Letter to Maurice Golding, 20 October 1964, RIBA Library, Specifications Panel Papers Box 1.
LaMarre, Thomas. 'Afterword: Humans and Machines', in Muriel Combes, *Gilbert Simondon and the Philosophy of the Transindividual*, trans. Thomas LaMarre, Cambridge, Massachusetts: The MIT Press, 2013.
Lambert, Léopold. 'For an Allagmatic Architecture: Introduction to the work of Gilbert Simondon' available at https://thefunambulist.net/architecture/simondon-episode-01-for-an-allagmatic-architecture-introduction-to-the-work-of-gilbert-simondon (accessed 25 April 2021).
Latour, Bruno. *Pandora's Hope: Essays on the reality of science studies*, Cambridge, Massachusetts: Harvard University Press, 1999.
Latour, Bruno. 'Why Has Critique Run out of Steam? From Matters of Fact to Matters of Concern' in *Critical Inquiry*, 30(2), 2004; 225–248.
Latour, Bruno. 'Mixing Humans and Nonhumans Together' in William Braham and Jonathan Hale (eds), *Rethinking Technology: A Reader in Architectural Theory*, London: Routledge, 2007.
Latour, Bruno. 'Which politics for which artifacts?' in *Domus*, June 2004, www.bruno-latour.fr/presse/presse_art/GB-06 per cent20DOMUS per cent2006-04.html (accessed 9 January 2009).
Leatherbarrow, David. 'Architecture's Unscripted Performance', in Kolarevic (ed.), *Performative Architecture*, New York: Spon Press, 2005.
Lecourt, Dominique. *Marxism and Epistemology: Bachelard, Canguilhem and Foucault*, trans. Ben Brewster, London: NLB, 1975.
Lefebvre, Pauline. '"What the Wood Wants to Do": Pragmatist speculations on a response-able architectural practice' in *Architectural Theory Review* 2.1, 2018; 24–41.
Light, Air and Simplicity: House at Gerrards Cross [1936], Daniell, Angela, director and producer, London: Momentum, 1997, available at RIBA library, VID0057.
Lloyd Thomas, Katie (ed.), *Material Matters: Architecture and Material Practice*, London: Routledge, 2007.
Lloyd Thomas, Katie. 'Jigging with Concrete: matter and form in the making of Wall One' in Alan Chandler and Remo Pedreschi (eds), *Fabricformwork*, London: RIBA Publications, 2007; 44–57.
Lloyd Thomas, Katie. 'Specifying Materials: language, matter and the conspiracy of muteness' in Marco Frascari, Jonathan Hale and Bradley Starkey (eds), *From* Models *to* Drawings, (London: Routledge, 2007); 242–252.
Lloyd Thomas, Katie. 'The excessive materiality of Stock Orchard Street: towards a feminist material practice' in Sarah Wigglesworth (ed.), *Around and About Stock Orchard Street*, London: Routledge, 2010; 118–131.
Lloyd Thomas, Katie. 'The Architect as Shopper: Women, electricity, building products and the interwar 'proprietary turn' in the UK' in Hélène Frichot, Catharina Gabrielsson, Helen Runting (eds), *Architecture and Feminisms: Economies, ecologies, technologies*, London: Routledge, 2018.
Lloyd Thomas, Katie. 'This Strange Interloper: building products and the emergence of the architect-shopper in 1930s' Britain' in Elizabeth Darling and Nathaniel R. Walker (eds), *Suffragette City: Women, Politics and the Built Environment*, London: Routledge, 2019.
Lloyd Thomas, Katie. 'Specifications: Writing Materials in architecture and philosophy' in *ARQ* 8, 3/4 (Dec 2004); 277–283.

Lloyd Thomas, Katie. 'Going Into the Mould: Process and materials in the architectural specification' in *Radical Philosophy*, 144, Jul/Aug 2007; 16–25.

Lloyd Thomas, Katie. 'Casting Operations and the Description of Process' in *Journal of Architecture* 20(3), April 2015; 430–444.

Lloyd Thomas, Katie and Nick Beech. 'Into the Hidden Abode: Architecture, Production, Process' in *Architecture and Culture* 3(3), November 2015; 271–279.

Lloyd Thomas, Katie and Tilo Amhoff. 'Writing Work: Changing Practices of Architectural Specification' in Peggy Deamer (ed.) *The Architect as Worker: Immaterial labor, the creative class, and the politics of design*, London: Bloomsbury, 2015.

Lloyd Thomas, Katie, Nick Beech and Tilo Amhoff (eds). *Industries of Architecture*, London: Routledge, 2015.

Locke, Sylvia, letter to Roderick + Innes and Co. Ltd., chartered surveyors, 16 April 1964, RIBA Library, Specifications Panel (of the technical information committee) Papers Box 1.

Lopes, João Marcos de Almeida. 'Architecture as ensemble: A matter of method', trans. Alice Fiuza, in Katie Lloyd Thomas, Nick Beech and Tilo Amhoff (eds), *Industries of Architecture*, London Routledge, 2016; 106–113.

Lyons, Arthur. *Materials for Architects & Builders*, Oxford: Elsevier Ltd., 2010.

Lyotard, Jean-Francois. *The Postmodern Condition: A Report on Knowledge*, trans. Geoff Bennington and Brian Massumi, Minneapolis: University of Minnesota Press, 1984.

Macey, Frank. *Specifications in Detail*, London: E. & F.N. Spon, 1898. 1904 edition reprinted, *Specifications in Detail*, Shaftesbury: Donhead, 2009.

Mackenzie, Adrian. *Transductions: Bodies and Machines at Speed*, London: Continuum, 2002.

Marx, Karl. *Capital: A Critique of Political Economy, Vol.I – The Process of Capitalist Production*, trans. Samuel Moore and Edward Aveling, (ed.) Ernest Untermann, Chicago: Charles H Kerr and Company, 1906.

Marx, Karl. *Capital Volume 1*, trans. Ben Fowkes, London: Penguin Books, 1990.

Massumi, Brian. *A User's Guide to Capitalism and Schizophrenia*, Cambridge, Massachusetts: The MIT Press, 1992.

Mazzilli-Daeschel, Stefano. 'Simondon and the Maker Movement' in *Culture, Theory and Critique* 60:3–4, 2019; 237–249.

McKenzie, Jon. *Perform or Else: From Discipline to Performance*, London: Routledge, 2001.

McMullin, Ernan (ed.). *The Concept of Matter in Greek and Medieval Philosophy*, Notre Dame: University of Notre Dame Press, 1965.

McMullin, Ernan. 'The Concept of Matter' in Ernan McMullin (ed.), *The Concept of Matter in Modern Philosophy*, Notre Dame: University of Notre Dame Press, 1978.

McVicar, Mhairi. *Precision in Architecture: Certainty, Ambiguity and Deviation*, London: Routledge, 2019.

Michelmore's Report, Panel meeting in 1963, RIBA Library, Specifications Panel Papers Box 1.

Moravánsky, Ákos. *Metamorphism: Material change in architecture*, Basel: Birkhäuser, 2018.

National Building Specification, London: RIBA Publications Ltd., 1973.

NBS Building, Newcastle: NBS, 3 Nov 2009.

Neimanis, Astrida. *Bodies of Water: Posthuman Feminist Phenomenology*, London: Bloomsbury, 2017.

Nietzsche, Friedrich. *Untimely Meditations*, trans. R.J. Hollingdale, Cambridge: Cambridge University Press, 1997.

'Olympia, 1936' in *The Builder*, Sept 4, 1936; 296.

Osman, Michael. *Modernism's Invisible Hand: Regulation and Architecture in America*, Minneapolis: University of Minnesota Press, 2018.

Osman, Michael. 'Specifying: The Generality of Clerical Labor' in Zeynep Çelik Alexander and John May (eds), *Design Technics: Archaeologies of architectural practice*, Minneapolis: University of Minnesota Press, 2020; 129–162.

Pallasmaa, Juhani. *The Eyes of the Skin: Architecture and the Senses*, Chichester: Wiley-Academy, 2005.

Piaget, Jean. *Judgment and Reasoning in the Child*, trans. Marjorie Walden, (first publ. 1928) London: Routledge, 2014.

Picon, Antoine. *Digital Culture in Architecture*, Basel: Birkhäuser, 2010.

Preston, Julieanna. *Performing Matter: Interior Surface and Feminist Actions*, Baunach: Spurbuchverlag, 2014.

Protevi, John. *Political Physics: Deleuze, Derrida and the Body Politic*, London: Athlone Press, 2001.

Powell, Christopher. *The British Building Industry since 1800: An Economic History*, London: E & FN Spon, 1996.

Puig de la Bellacasa, Maria. *Matters of Care: Speculative Ethics in More Than Human Worlds*, Minneapolis: University of Minnesota Press, 2017.

Rawes, Peg (ed.). *Relational Architectural Ecologies: Architecture, nature and subjectivity*, London: Routledge, 2013.

Ray-Jones, Allen report, Panel meeting notes, 1963 (RIBA Library, Specifications Panel Papers Box 1).

Reiser + Umemoto. *Atlas of Novel Tectonics*, New York: Princeton Architectural Press, 2006.

Roberts, Tom. 'Thinking Technology for the Anthropocene: Encountering 3D printing through the philosophy of Gilbert Simondon' in *Cultural Geographies* 24(4), 2017; 539–554.

Rosen, Harold J. 'The Future of Specifications Writing – Part II' in *Progressive Architecture*, September 1969, 206.

Ross, Liam. 'On contradictory regulations' in *Architectural Research Quarterly* 16(3), 2012; 205–209.

Ross, Liam. 'Regulatory spaces, physical and metaphorical: on the legal and spatial occupation of fire-safety regulation' in Lloyd Thomas, Katie, Nick Beech and Tilo Amhoff (eds). *Industries of Architecture*, London: Routledge, 2015; 235–244.

Roux, Hannah le, Gabrielle Hecht. 'Bad Earth', www.e-flux.com/architecture/accumulation/345106/bad-earth/ (accessed 3 May 2021).

Rowe, Colin and Robert Slutzky, 'Transparency: Literal and Phenomenal' in *Perspecta* 8, 1963.

Ruskin, John. *Stones of Venice*, Jan Morris (ed.), London: Faber, 1991.

Salter, Peter. *TS: Intuition and Process*, London: The Architectural Association, 1989.

Schumann, Nick. 'The Watchful Protector of Quality', *Architects Journal*, 28 October 2013, http://schumannconsultltd.com/blog/watchful-protector-quality-architects-journal-october-column/#sthash.1IkYqEIE.dpbs (accessed 2 November 2013).
Sereda, P.J. 'Performance of Building Materials', in *Canadian Building Digest 115*, July 1969. http://irc.nrc-cnrc.gc.ca/pubs/cbd/cbd115_e.html (accessed 10 June 2009).
Sharr, Adam. *Modern Architecture: A very short introduction*, Oxford: Oxford University Press, 2018.
Shonfield, Katherine. *Walls Have Feelings: Architecture, Film and the City*, London: Routledge, 2000.
Shonfield, Katherine, Rosa Ainley and Adrian Dannatt (eds). *This Is What We Do: A Muf Manual*, London: Batsford, 2001.
Shotton, Elisabeth. 'Material Imprecision' in Katie Lloyd Thomas (ed.), *Material Matters*, London: Routledge, 2007.
Simondon, Gilbert. *L'individu et sa gènese physico-biologique*, Paris: Presses Universitaires de France, 1964.
Simondon, Gilbert. *Individuation in Light of Notions of Form and Information*, trans. Taylor Adkins, Minneapolis: University of Minnesota Press, 2020.
Simondon, Gilbert. *Du mode d'existence des object techniques*, Paris: Aubier, 2001.
Simondon, Gilbert. *L'Invention Dans Les Techniques: Cours et conferences*, (ed.) Jean-Yves Chateau, Paris: Éditions du Seuil, 2005.
Simondon, Gilbert. *L'individuation à la lumière des notions de forme et d'information*, Grenoble: Editions Jérôme Millon, 2006.
Simondon, Gilbert. *L'individuation psychique et collective*, Paris: Aubier, 2007.
Simondon, Gilbert. *On the Mode of Existence of Technical Objects*, trans. Cécile Malaspina and John Rogove, Minneapolis: Univocal Publishing, 2017.
Simondon, Gilbert. 'The Genesis of the Individual', trans. Mark Cohen and Sanford Kwinter, in Sanford Kwinter, Jonathan Crary (eds), *Incorporations*, New York: Zone Books, 1992.
Simondon, Gilbert. 'The Physico-biological Genesis of the Individual'. (Chapter 1, parts I and II only), trans. Taylor Adkins as and posted at http://fractalontology.wordpress.com /2007/ 10/03/ translation-simondon-and-the-physico-biological-genesis-of-the-individual/ (accessed 4 December 2008).
Simondon, Gilbert. 'Individuation of Perceptive Units and Signification' (Chapter 1 only), trans. Taylor Adkins and posted at http://speculativeheresy.wordpress.com/2008/10/06/translation-chapter-1-of-simondons-psychic-and-collective-individuation/ (accessed 30 April 2010).
Simondon, Gilbert. 'On Techno-Aesthetics', trans. Arne de Boever, in *Parrhesia* 14, www.parrhesiajournal.org/parrhesia14/parrhesia 14_Simondon/pdf, (accessed 18 July 2013).
Smith, Cyril Stanley. *A Search for Structure: Selected Essays on Science, Art and History*, Cambridge, Massachusetts: MIT Press, 1982.
Specification Vol 1, 1898; 1964; 1935.
Spuybroek, Lars. *NOX: Machining Architecture*, London: Thames and Hudson, 2004.
Stiegler, Bernard. *Technics and Time, Vol.1: The Fault of Epimetheus*, trans. Richard Beardsworth and George Collins, Stanford: Stanford University Press, 1998.

Stroud, Dorothy. *George Dance, Architect 1745–1825*, London: Faber and Faber, 1971.
Suare, Ruben. 'Innovation Through Accountability in the Design and Manufacturing of Material Effects' in Branko Kolarevic and Kevin Klinger (eds), *Manufacturing Material Effects: Rethinking Design and Making in Architecture*, London: Routledge, 2008.
Summerson, John. *Georgian London*, London: Pleiades Books, 1945.
Swenarton Mark. *Artisans and Architects: The Ruskinian tradition in architectural thought*, Basingstoke: Macmillan, 1989.
Theophilius. *On Divers Arts*, trans. and introduced by John G. Hawthorne and Cyril Stanley Smith, New York: Dover Publications, 1979.
Thompson, Emily. *The Soundscape of Modernity: Architectural Acoustics and the Culture of Listening in America, 1900–1933*, Cambridge, Massachusetts: MIT Press, 2004.
Till, Jeremy and Sarah Wigglesworth. 'The Future is Hairy', in Jonathan Hill (ed.), *Architecture: The Subject is Matter*, London: Routledge, 2001. Toscano, Alberto. 'Technical Culture and the Limits of Interaction: A Note on Simondon' in Joke Brouwer and Arjen Mulder (eds), *Interact of Die!*, Rotterdam: V2_Publishing/NAi Publishers, 2007.
Toscano, Alberto. *The Theatre of Production: Philosophy and Individuation between Kant and Deleuze,* Basingstoke: Palgrave Macmillan, 2006.
Virno, Paolo, 'Reading Gilbert Simondon: Transindivuality, technical activity and reification' an interview with Jun Fujita Hirose, trans. Matteo Mandarani, *Radical Philosophy* No.136, March/April 2006.
Voss, Daniela. 'Invention and Capture: a critique of Simondon' in *Culture, Theory and Critique* 60, 3–4; 279–299.
Wall, Christine. *An Architecture of Parts: Architects, Building Workers and Industrialisation in Britain 1940–1970*, London: Routledge, 2013.
Westerlind, Helena. 'Concrete Deposition: Choreographing Flow', in Bob Sheil, Mette Ramsgaard Thomsen, Martin Tamke and Sean Hanna (eds), *Design Transactions: Rethinking information modeling for a new material age*, London: UCL Press, 2020, 196–7.
Whitehead, Alfred North. *Science and the Modern World*, New York: Mentor Books, 1964.
Yaneva, Albena. *Mapping Controversies in Architecture*, Abingdon: Routledge, 2016.
Yorke, F.R.S., 'Preface' in *Specification* 1936; iii.

Specifications

Documents are listed in date order, followed by the title / description, architect or contractor's name if given, and then by source, whether citation in a book, archival document or given to me by an architectural practice.

(1734–6) Articles of Agreement for a town house for Sir William Heathcote at St James Square, London. Architect – Henry Flitcroft, RIBA Archives, HeW/1/1/2.

(1769) Specification for Newgate Gaol and the Sessions-House at the Old Bailey, London, architect – George Dance the Younger. RIBA Library EW No. 790.
(1770) Particulars for a bakehouse, dwelling and lofts in parish of St Anne, Middlesex, contractor – Robert Wilson, client – John Steinmetz. RIBA Archives WiR/1.
(1784) The particulars and estimates of the several works for Tendring Hall, architect – John Soane. Soane Museum, SM 28/3/1a/1–6.
(1791–2) Description of the several works to be done in erecting a new chapel and wing buildings in St James Burial Ground, Tottenham Court Road architect – Thomas Hardwick, RIBA Archives AC/HAR/Add/2.
(1819) Specification for the Erection of a School house at Hammersmith, architect – Charles Fowler. RIBA Archives, FOC/1/1.
(1833–43) *Specifications*, architect – David Mocatta, (RIBA Archives, MoD/1/1).
(1835) Specification of the work, and particulars of the materials to be used in the erection of an asylum for the blind, Bristol, architects – Rickman and Hussey, RIBA library, EW 2772.
(1843) Specification of sundry works required in rebuilding the house no.1 Princes Street Leicester Square, architect – Thomas Little, (RIBA Archives LIT/1/1 + 2).
(1879) Specification of works for a 'Gothic' villa at Godden Green, near Sevenoaks, Kent for Mr Usborne. Architect – Joseph Fogerty. RIBA Archives, FoJo/1/1.
(1909) Specification for Shops/Offices, Oxford St./Rathbone Pl., London, architects – Holden and Adams. RIBA Archives, AHP/1/18/1.
(1934) Specification of Works for a House at Farnham Common, Bucks. Architect – Val Harding with Tecton. RIBA Archives, SaG/17/3.
(1934–5) Specification of Works for a smoked sausage manufactory, 5 Fairhazel Gardens, NW6. Architect – Gerhard Rosenberg. RIBA Archives, SaG/17/7.
(1936) Specification for house at Hedgerley Lane, Gerrards Cross, Bucks, architect – Elisabeth Benjamin, (RIBA Archives, SaG/9/4).
(1940–41) Specification for Air Raid Shelter for Bedales, Hants., architects – Erno Golfinger with Mary Crowley. RIBA Archives, GolER/409/5.
(1961) Specification for the Works at the Elfrida Rathbone School for the Educationally Subnormal, Camberwell. Architect – John Bancroft with London County Council. RIBA Archives, LCC/AD/1.
(2004) *National Building Specification: Standard Version (Update 38)*. London: RIBA Enterprises Ltd. 2004.
(2004) Specification for Sure Start, St Anne's Ward, Colchester 2004, architect – DSDHA, Courtesy DSDHA.
(2005) Specification: Project Title: Sainsbury's Maidenhead. For Construction, architect – Chetwoods®. Courtesy Chetwoods®.
(2005) Specification for the Q . . . t Building, prepared by Schumann Smith for . . . Architects. Courtesy Schumann Smith.
(2005) Specification for Greenhouse and Summerhouse at Richmond Terrace Gardens, architects – Caroe & Partners. Courtesy René Tobe.
(2006) Specification, The H . . . y Building, prepared by Schumann Smith for . . . Architects. Courtesy Schumann Smith.

(2006) Architectural Specification for Parkside and Blossom Square Kiosks, Potters' Fields Park, architect – DSDHA. Courtesy DSDHA.
(2006) Specification for Refurbishment of Scroope Terrace & New Studio Building, University of Cambridge Faculty of Architecture, architects – Mole Architects. Courtesy Mole Architects.
(2007) Specification (in production) for a group of office buildings, Coventry, architect – Allies and Morrison. Courtesy: Allies and Morrison.

Index

The letter *f* following an entry indicates a page that includes a figure.

Addington, Michelle and Schodek, Daniel 135, 142, 158
Akrich, Madeleine 149–50
Alice Tully Hall (Diller & Scofido + Renfro) 173–4
alienation 195–6, 198
allagmatic theory 81
Allies and Morrison 140, 145
Allot, Tony 77, 120
Archaeology of Knowledge, The (Foucault, Michel) 29
architects 44, 45, 60, 197–8, 199–200
    hylomorphism 102
    proprietary specification 64–6
    quality control 186–7
    stand design 59–60
architectural action 118
architectural drawings 1–2, 8
architectural fabrication 101–2
'Architectural Room Specifications' (Burroughs, Brady) 202
architecture
    aesthetics 226 n.28
    economic, social and regulatory considerations 166–7
    environmental concerns 168–9, 220 n.16
    industrial and technical considerations 167–8
    Lopes, João Marcos de Almeida 200–1
    production and mobilization considerations 168, 181, 182, 184–7, 188
Aristotle 2, 11, 68, 69, 71
    Metaphysics 66–7, 79–80
Armstrong, Rachel
    Vibrant Architecture 2

Arnhoff, T., Beech, N. and Lloyd Thomas, K.
    Further Reading Required seminar 6
    Industries of Architecture 6
Arnhoff, Tilo 37
Art of the Metalworker, The (Theophilius) 85–6
artisans 91, 200
Arts Decoratifs Aujourd Hui, Les (Le Corbusier) 62

Bachelard, Gaston 80, 104
'Bad Earth' (le Roux, Hannah and Hecht, Gabrielle) 169
Ball, Michael 7
Ballard-Bell, Victoria
    Materials for Architectural Design 57, 58
Bamboo Canopy (nArchitects) 57
Bancroft, John 74
Barnstone, Deborah Ascher 136
barrier system requirements 138, 139
Barry, Andrew 162–3, 165
    'Pharmaceutical Matters: The Invention of Informed Materials' 156–7, 158
Barthélémy, Jean-Hughes 21
Barthes, Roland 78–9, 80–1
Bartholomew, Alfred
    Specifications for Practical Architecture 40, 42
Bedales air raid shelter 115
Benjamin, Andrew 167
Benjamin, Elisabeth
    East Wall, Gerrards Cross 61–4
Bennett, Jane
    Vibrant Matter 2

Bensaude-Vincent, Bernadette and Stengers, Isabelle
History of Chemistry 142–3, 156, 157, 158
Bible, the 51
books 183–4
brand names 60–1, 65
Brassington, Kevan 123
bricks 11, 12, 13, 189, 201
    brick-making 197
    form-taking 18, 86
    preliminary operations 110
    process 73
    Simondon, Gilbert 73, 86
    veritable relations 19
Bristol Asylum for the Blind 137
brutalism 93
Buchanan, Peter 136
builder techniques 38, 178
building
    cultures of 8
    extra-constructional 185
    modes 217 n.46
    techniques 38, 178
    unconventional 175–81, 184–5
Building Control 179
buildings
    behaviours 118–19
    structural, thermal, material stability 118
Bundeshaus, Bonn (Schwippert, Hans) 136, 137, 142
Burroughs, Brady
    'Architectural Room Specifications' 202
Button, David and Pye, Brian
Glass in Building 134–5, 142, 144

Cambridge, school of architecture extension (Mole Architects) 115–16, 118
Capital 1 (Marx, Karl) 195–6, 200–1
capitalism 195–6
Caroe and Partners Architects
    primitive hut 175
Carter, John 77, 122–3
cast stone 121–2
    wall dressings 126–8

Centro de Trabalhos para o Ambiente Habitado/FACTORY – Work Centre for Living Space (USINA CTAH) 201
Chandler, Alan and Pedreschi, Remo 'Wall One' 87–90, 198
Chandler, Ian 113
charring 175–7
chemistry 142–3, 156–8, 165–6
Chetwoods@
    Sainsbury's curtain walling system 126–8, 129
clauses 9–10, 47–8, 50, 185 *see also* naming *and* performance clauses *and* process-based clauses
    extra-physical 108
    Foucault, Michel 32, 53
    Newgate Gaol, Old Bailey 54
    selection 14
    Sessions House, Old Bailey 54
    variety of forms 17, 20, 24, 47–8, 50, 181–2, 187
clay 11, 12, 13, 189
    clay/mould system 110, 159, 160, 161
    form-setting 108
    form-taking 18, 86
    moulding 92
    preliminary operations 106–7, 110
    process 73
    Simondon, Gilbert 73, 83, 86, 106–7, 108
    veritable relations 19
closed construction methods 47
Colaço-Osorio, Arnold 61, 62–4
collectivity 193, 201
colonialism 169
compatibility 169–70, 187–9
concrete 11, 168–9, 170
    matterization of 104
    preliminary operations 107–8
    as a process 184
    timber grained 98–100
Concrete and Culture (Forty, Adrian) 168, 184
concrete casting 19, 74, 76
    preliminary operations 108–9

INDEX

process-based clauses 19, 74, 76, 93, 96–103
concrete fabric-formworking 87–90
concretization 199
constant materials 105, 106, 121
constitutive seam 166–82
construction
   cultures of 8
   extra-constructional 185
   modes 217 n.46
   techniques 38, 178
   unconventional 175–81, 184–5
Construction Materials Manual (publication) 57
contracting in gross 30, 35
contracts 49–50, 52
Copromo, Osaco 201
costs 42
   Donaldson, Thomas 39
   Specification (journal) 42
Cottrell & Vermeulen Architecture 177
craft 215 n.6
craft manuals 72
Critique of Pure Reason, The (Kant, Immanuel) 66
crystal formation 110, 160–3

Dance, George the younger
   Newgate Gaol and the Sessions House, Old Bailey. *See* Newgate Gaol, Old Bailey *and* Sessions House, Old Bailey
De Architectura (Vitruvius) 72
DeLanda, Manuel 87, 105–6
delegation 144
Deleuze, Gilles and Guattari, Félix
   metalworking 92–3
   Thousand Plateaus, A 90–1
Description of the several works to be done in erecting a new chapel and wing buildings on St James Burial Ground, Tottenham Court Road (Hardwick, Thomas) 71–2, 73, 75
descriptive specification 48, 49, 171
Design and Build contract 30
design intent 173 *see also* visual intent

Digital Culture in Architecture (Picon, Antoine) 119
Digital Grotesque project (Hansmeyer, Michael and Dillenburger, Benjain) 199
digital technology 199
Diller & Scofido + Renfro
   Alice Tully Hall 173–4
diversity 4–5
Donaldson, Thomas
   Handbook of Specifications 39, 40, 42
drawings. *See* architectural drawings
DSDHA
   Pottersfield Park Pavilion 137–8, 175–7
   Sure Start, St Anne's Ward, Colchester 56–7
Du mode d'existence des object techniques (Simondon, Gilbert) 10, 12, 15, 17, 21, 25, 189–90, 193–5
dynamic operations 86–103, 154

East Wall, Gerrards Cross (Benjamin, Elisabeth) 61–4
ecological concerns 168, 181, 220 n.16
economic considerations 166–7
Egypt End, Farnham Common (Harding, Val) 55, 74, 76, 93, 96–8, 187
Eisenman, Peter 168
   'Real and English: The Destruction of the Box' 168
electronic engineering 21
Elfrida Rathbone School for the Educationally Subnormal 43, 74, 75, 76
   paintwork specification 185–7
   preliminary operations 187
   process-based description 93–6, 98–100
ensembles 163–4
   preliminary operations 107, 108–9
   technical ensembles 192, 200
environmental concerns 168–9, 181, 220 n.16

equipment 15, 17, 129–34
  thingness of 15, 131
evolution (of technical objects) 190–2, 199
extra-constructional building 185
extra-physical clauses 108
extra-physical processes 20

fabric formwork 87–90, 198
factor relationships 17–18
Farocki, Harun
  In Comparison 197
Fernandez, John
  Material Architecture: Emergent materials 138, 139–40
Ferro, Sérgio 4, 168–9
Fore Street Bakehouse, dwelling and lofts 34–5
form 2, 3 *see also* visual intent
  fabric formwork 87–90, 198
  Heidegger, Martin 15
form/matter schema
  architects 102
  Aristotle 66–7, 68, 69, 61, 79–80
  Bachelard, Gaston 80
  criticism of 86–93, 130
  DeLanda, Manuel 87
  Deleuze, Gilles and Guattari, Félix 90–3
  Egypt End, Farnham Common 98
  forgetting of process 78–86
  Heidegger, Martin 15–16, 130
  Lopes, João Marcos de Almeida 200–1
  matterization 104
  Reiser + Umemoto 102
  Simondon, Gilbert 11, 14–15, 73, 81–3, 101–4, 130
  Table 2/3 67–70, 71
  3D printing 199
  universality 103–4
  work 194
form-taking 12–13, 18, 154, 189
formless materials 70, 71
forms of clauses 47
  classification 47
Forty, Adrian
  Concrete and Culture 168, 184

Foster, Norman
  Reichstag, Berlin 136–7
Foucault, Michel 32, 53
  Archaeology of Knowledge, The 29
Fowler, Charles
  Specification for the Erection of a School house at Hammersmith 75
Fundamental Concepts of Metaphysics, The (Heidegger, Martin) 132–3
funicity 1
Further Reading Required seminar (Amhoff, T., Beech, N. and Lloyd Thomas, K.) 6

Gaetano, Sergio de 149
Gehry, Frank 1
  Santa Monica house 66–7
Gelder, John 38
  Specifying Architecture: A Guide to Professional Practice 6, 47–8
Genesis 6. 14–16 51
Giedion, Sigfried
  Mechanization Takes Command 27
Gille, Bertrand
  History of Techniques 72, 197
glass
  cleaning 171–2
  coatings 141–2f
  delegation, human to non-human 144
  extra-physical processes 19–20
  heat transmission 147–8
  impact testing 145–6f
  naming 54
  performance-engineered 16, 134–5, 141–2, 144–6, 147–50, 187
  Pilkington 134–5, 141, 142f, 143, 144–5, 148
  security 149
  specifications 137–8, 140
  transparency (political) 136–7, 149
Glass in Building (Button, David and Pye, Brian) 134–5, 142, 144
goals, 16, 20, 128, 145, 150–1
  Latour, Bruno 143, 144, 145

INDEX

gopher wood 51
gothic villa, Gooden Green for Mr Usborne 16, 114
government policies 35–7
gross tendering 38–9
Gucklhupf, Mondsee (Worndl, Hans Peter) 57
guild system 6, 7

H. . .y Building (Schuman Smith) 31, 49, 138–9
Handbook of Specifications (Donaldson, Thomas) 39, 40, 42
Hansmeyer, Michael and Dillenburger, Benjain
  Digital Grotesque project 199
Hanson, Brian 215 n.6
hardcore 170
Harding, Val
  Egypt End, Farnham Common 55, 74, 76, 93, 96–8
Hardwick, Thomas
  Description of the several works to be done in erecting a new chapel and wing buildings on St James Burial Ground, Tottenham Court Road 71–2, 73, 75
HBBE (Hub for Biotechnology in the Built Environment), Newcastle University 198
Heidegger, Martin 17, 129, 130–3
  Fundamental Concepts of Metaphysics, The 132–3
  'Origin of the Work of Art, The' 15–16, 130–2, 133
History of Chemistry (Bensaude-Vincent, Bernadette and Stengers, Isabelle) 142–3, 156, 157, 158
History of Techniques (Gille, Bertrand) 72, 197
homogenous materials 105–6, 121
Houses & Materials (publication) 57
housing 201
Hub for Biotechnology in the Built Environment (HBBE), Newcastle University 198

human to non-human, delegation 143–4
humans
  technical objects, relations with 189–90, 191–7, 198, 200
hylomorphism (form/matter schema) 2, 3, 78–86
  architects 102
  Aristotle 66–7, 68, 69, 61, 79–80
  Bachelard, Gaston 80
  criticism of 86–93, 130
  DeLanda, Manuel 87
  Deleuze, Gilles and Guattari, Félix 90–3
  Egypt End, Farnham Common 98
  forgetting of process 78–86
  Heidegger, Martin 15–16, 130
  Lopes, João Marcos de Almeida 200–1
  matterization 104
  Reiser + Umemoto 102
  Simondon, Gilbert 11, 14–15, 73, 81–3, 101–4, 130
  Table 2/3 67–70, 71
  3D printing 199
  universality 103–4
  work 194

implicit forms 91–2, 105
'(Im) Possible Instructions: Inscribing use-value in the architectural design process' (Svenningsen Kajita, Heidi) 202
In Comparison (Farocki, Harun) 197
indicative material 20, 52, 172–3, 175–6
L'individu et sa genèse physico-biologique (Simondon, Gilbert) 10, 11, 14, 20–1, 183
  crystal formation 160
  process 73, 81, 82, 86
  wet clay brick 11, 24, 73, 82, 84, 189
  work 194–5
individuation 18, 81–4, 130, 153, 155
  complete system of 153–5, 159–66
  crystal formation 110, 160–3
  ensembles 163–4

information 158–9, 161–5
invention 164–5
Lopes, João Marcos de Almeida 201
Mackenzie, Adrian 158–9
mediation 159–60, 163–5
metastability 18–19, 110, 153, 158, 160, 164
photosynthesis 163
physical mode 155
relation, conditions for 187–8
Simondon, Gilbert 10–12, 15, 18–21, 24, 81–4, 158–60
systems of 20–1
transduction 21–3
transindividuation 189, 193
veritable relations 17, 19
L'individuation à la lumière des notions de forme et d'information (Simondon, Gilbert) 10, 14, 81
'Individuation and Invention' (Simondon, Gilbert) 189, 190, 193–4, 195
Individuation in Light of Notions of Form and Information (Simondon, Gilbert) 10
L'individuation psychique et collective (Simondon, Gilbert) 11, 163–4, 189
industrialization 200
Industries of Architecture (Lloyd Thomas, K, Amhoff, T. and Beech, N.) 6
information 158–9, 161–5
informed material 156–60
International Style 93, 168
Intuition and Process (Salter, Peter)
invention 164–5, 190–2, 198–9, 201–2
L'Invention Dans les Techniques: Cours et conferences (Simondon, Gilbert) 21, 164, 165
inventive relations 197–203
isotropic materials 105

Jameson, Frederic 166–7
Jarzombek, Mark
'Quadrivium Industrial Complex, The' 168

Kant, Immanuel
Critique of Pure Reason, The 66
Knightly Brown, Andrew 28
Kolarevic, Branko 119
Performative Architecture 117

labour 4, 106, 193–7, 201
Latour, Bruno 16, 143–4, 145
pre-inscription 145
prescription 149–50
translation concept 145, 151
LCC (London County Council) 42–3
'Specification for the Elfrida Rathbone School'. *See* Elfrida Rathbone School for the Educationally Subnormal
Le Corbusier
Arts Decoratifs Aujourd Hui, Les 62
Sainte Marie de La Tourette 101
le Roux, Hannah and Hecht, Gabrielle 'Bad Earth' 169
Leatherbarrow, David 117–19
Lecourt, Dominique 104
Lefebvre, Pauline 90
Leicester Engineering Building (Stirling, James) 168
Levi, Primo
Periodic Table, The 85
Wrench, The 85
Little, Thomas
specification for house on Princes Street 61
load transfer requirements 138–9
London 35–7 *see also* LCC
Elfrida Rathbone School for the Educationally Subnormal. *See* Elfrida Rathbone School for the Educationally Subnormal
Fore Street Bakehouse, dwelling and lofts 34–5
Hammersmith schoolhouse 75
London Fever Hospital 61, 75
Marylebone Street houses 54
Newgate Gaol and the Sessions House, Old Bailey. *See* Newgate Gaol, Old Bailey *and* Sessions House, Old Bailey

St James Burial Ground,
Tottenham Court Road 71–2,
73, 75
St James Square town house 31,
32–4
London County Council (LCC). See
LCC
London Fever Hospital (Mocatta,
David) 61, 75
Lopes, João Marcos de Almeida
200–1
Lyons, Arthur
Materials for Architects & Builders
57–8
Lyotard, Jean-Francois 119

Macey, Frank
Specifications in Detail 42
machines 194–7
Mackenzie, Adrian 158–9
McKenzie, Jon
Perform or Else 119
McVicar, Mhairi
*Precision in Architecture: Certainty,
Ambiguity and Deviation* 7
Mappae Clavicula 72
Marx, Karl 13–14, 106, 195–6
Capital 1 195–6, 200–1
Marylebone Street houses (Mocatta,
David) 54
Massumi, Brian 109
*User's Guide to Capitalism and
Schizophrenia, A* 110
*Material Architecture: Emergent
materials* (Fernandez, John)
138, 139–40
material turn, the 168
materials, overlooked 1
Materials for Architects & Builders
(Lyons, Arthur) 57–8
Materials for Architectural Design
(Ballard-Bell, Victoria) 57, 58
materials science 16, 115
Material World (publications) 57
matterization 104–6
measurement 115–16, 170
Mechanization Takes Command
(Giedion, Sigfried) 27
mediation 159–60, 163–5

metalworking 92–3
Metamorphism (Moravánsky, Ákos)
167
Metaphysics (Aristotle) 66–7, 79–80
metastability 18–19, 110, 153, 158,
160, 164 *see also* preindividual
state *and* preliminary operations
Michelmore, Mr 47–8
'Report to the Interim
Specifications Panel' 76–7
mobilization 168, 181, 182, 184–7,
188
Mocatta, David 40
Specification for building a
workshop 107
specification for the London Fever
Hospital 61, 75
Specification for Two houses on
Marylebone Street 54
Specifications 40, 42
modernism 93, 100
Mole Architects
school of architecture extension
Cambridge 115–16, 118
Moravánsky, Ákos
Metamorphism 167
mortar 71–2, 73
Fowler, Charles 75
preliminary operations 107, 108
moulding 83–4, 91, 92

naming materials 14, 51–3, 57–8, 66,
185, 209 n.8
brand names 60–1, 65
materials as varieties of matter
66–70
proprietary specification 59–65
Table 2/3 67–70, 71
timber 51, 53–9
nArchitects
Bamboo Canopy 57
Nash, John 38–9
National Building Specification (NBS).
See NBS
natural stone 49, 120–1, 172
NBS (National Building Specification)
31, 42, 44–7, 77–8, 181
named materials 67
performance specification 116, 122

promotion 202
proprietary specification 60
stone 121–2
unconventional construction 175
Newgate Gaol, Old Bailey (Dance, George the Younger) 35, 36f, 37, 38, 39
    naming 54
    painter's specification 39
Noah 51
NOX
    'Oblique WTC' 87
    'Soft Office' 87

'Oblique WTC' (NOX) 87
off-site manufacture 76
On Divers Arts (Theophilius) 72, 85
'On Techno-Aesthetics' (Simondon, Gilbert) 101
On the Mode of Existence of Technical Objects (Simondon, Gilbert) 10, 12, 84
ontogenetic approach 12–15, 187–8
open construction methods 47
open specifications 9, 31, 113–14, 171, 174
operation 81
'Origin of the Work of Art, The' (Heidegger, Martin) 15, 130–2, 133

paint 185–7
Parker's Roman Cement 61
Perform or Else (McKenzie, Jon) 119
performance 117–19, 188 *see also* performance clauses
    evaluating 123f–4f
    glass 134–5, 136–8, 140, 141–2f, 144–6, 147–50
    material as equipment 125–34
    performance matrix 140–1, 145
    performance specification 48, 49, 77, 113–25, 150, 153, 154, 188
    performance-engineered glass 16, 134–5, 141–2, 144–6, 147–50, 187
    performance-engineered materials 6, 134–52, 159
    pre-inscription 145
    predictions 147
    resistance to 49, 120–1, 172
    statutory mechanisms 179
    unconventional materials 179–81, 185
performance clauses 9, 14, 16, 52, 138, 153, 185
    acoustic, thermal and visual 140
    barrier system requirements 138, 139
    design intent 173–4
    extra-physical processes 20
    factor relationships 17–18
    load transfer requirements 138–9
    preparation 182
performance-engineered materials 16, 134–52
performance matrix 140–1, 145
'Performance of Building Materials' (Sereda, P. J.) 147
Performative Architecture (Kolarevic, Branko) 117
Periodic Table, The (Levi, Primo) 85
pharmaceutical industry 156–7, 158, 165–6
'Pharmaceutical Matters: The Invention of Informed Materials' (Barry, Andrew) 156–7, 158
phenomenology 167
philosophy 11–12
photography 165
photosynthesis 163
Picon, Antoine
    Digital Culture in Architecture 119
Pilkington 134–5, 141, 142f, 143, 144–5, 148
plastic 78–9, 80–1
plastic by preparation 103–11
Polaroid Land 165
Political Physics (Protevi, John) 11, 101–2
Portland stone 120–1, 172
Pottersfield Park Pavilion (DSDHA) 137–8, 175–7
power 44
pre-inscription 145
pre-technical objects 190

Precision in Architecture: Certainty, Ambiguity and Deviation (McVicar, Mhairi) 7
preindividual state 12, 18–19 *see also* metastability *and* preliminary operations
preliminary operations 13–14, 73, 74, 83–4, 103–11, 184–9 *see also* preparations
   books 183–4
   communication 19
   individuation 125, 152
   performance specification 124–5
   process 73, 74, 103–11, 154
   Simondon, Gilbert 13, 14, 73, 74, 83–4, 103–11, 125, 152
preparations 83–6, 182 *see also* preliminary operations
prescription 15, 52, 101, 131–3
   Heidegger, Martin 132
   Latour, Bruno 149–50
   prescriptive specification 48, 49, 137–8
   Schumann Smith 113, 132
Preston, Julieanna 3
primitive hut (Caroe and Partners Architects) 175
Princes Street house 61
principle of individuation 82
process 71–4 *see also* process-based clauses
   dynamic operations 86–103
   forgetting of process 78–86
   hylomorphism 78–86
   preliminary operations 73, 74, 103–11, 154
   rendered plastic by preparation 103–11
   Simondon, Gilbert 14–15, 73, 74, 86
Process and Reality (Whitehead, Alfred North) 157–8
process-based clauses 9, 14, 17, 52, 71–8, 134, 183, 185
   cleaning 171
   concrete casting 19, 92, 96–103
   Egypt End, Farnham Common 96–8

Elfrida Rathbone School specification 93–6, 98–100, 185–7
NBS 46–7
preparation 182
replacing 43–4
Simondon, Gilbert 154
process theory 14–15
production 79–80, 195, 197, 200
properties 115, 121, 123–5
proprietary specification 52–3, 59–65
Protevi, John 101–2
Political Physics 11, 101–2
putty 169–70

'Quadrivium Industrial Complex, The' (Jarzombek, Mark) 168
quality control 186–7
quantity surveyors 44

Ray-Jones, Allen 44
'Real and English: The Destruction of the Box' (Eisenman, Peter) 168
recipes 72, 84–5
   recipe-based specifications 75
Reichstag, Berlin (Foster, Norman) 136–7, 138
Reiser + Umemoto 102
relation, conditions for 187–8
'Report to the Interim Specifications Panel' (Michelmore, Mr) 76–7
RIBA (Royal Institute of British Architects) 31
   NBS 31, 44–6, 77
   Specifications Panel 43–4
Richard Rogers and Partners
   Welsh Assembly Building 149
road boulder 164–5, 193
Roberts, Tom 199
Rosen, Harold 116
Rosenberg's sausage factory 114

Sainsbury's curtain walling system (Chetwoods@) 126–8, 129
Sainte Marie de La Tourette (Le Corbusier) 101
Salter, Peter
   Intuition and Process 13
sandbags 177–9, 181

INDEX

Santa Monica house (Gehry, Frank) 166–7
Sarah Wigglesworth Architects (SWA)
　Stock Orchard Street straw bale house 177–9, 180f, 181, 184–5
Schuman Smith 28–9, 30, 48–50, 138, 181
　design intent 173
　indicative material 20, 52, 172–3
　naming 52
　open specifications 31, 113, 171, 174
　performance specification 116
　process-based clauses 171–2
　'Specification for the H. . .y Building' 31, 49, 138–9
Schwippert, Hans
　Bundeshaus, Bonn 136, 137, 142
science 157–8
　chemistry 142–3, 156–8, 165–6
　materials science 16, 115
seed (crystal) 110, 160–1
Sereda, P. J.
　'Performance of Building Materials' 147
Sessions House, Old Bailey (Dance, George the Younger) 35, 36f, 38
　carpenter's specification 61, 75
　glass specification 137
SfB classification system 45–6, 67
Simondon, Gilbert 2, 5, 10, 81
　allagmatic theory 81
　architectural fabrication 101
　Barthélémy, Jean-Hughes 21
　'complete system' 153–5, 159–66
　concretization 199
　crystal formation 110, 160–3
　Du mode d'existence des objects techniques 10, 12, 15, 17, 21, 25, 189–90 , 193–5
　dynamic operations 86–103
　'going into the mould' 83, 184, 189, 198, 202–3
　hylomorphism 11, 14–15, 73, 81–3, 101–4, 130
　implicit forms 91–2
　L'individu et sa genèse physico-biologique. See *L'individu et sa genèse physic-biologique*

individuation. See individuation
L'individuation à la lumière des notions de forme et d'information 10, 14, 81
'Individuation and Invention' 189, 190, 193–4, 195
Individuation in Light of Notions of Form and Information 10
L'individuation psychique et collective 11, 163–4, 189
information 158–9
invention 164–5, 198–9
L'Invention Dans les Techniques: Cours et conferences 21, 164, 165
Lopes, João Marcos de Almeida 200–1
mediation 159–60, 163–5
moulding 83–4, 91
'On Techno-Aesthetics' 101
On the Mode of Existence of Technical Objects 10, 12, 84
plastic by preparation 103–11
Polaroid Land 165
preindividual state 12, 18–19
preliminary operations 13, 14, 73, 74, 83–4, 103–11, 125, 152
process theory 14–15, 73, 74, 86
road boulder 164–5
system 153–5, 159–66
technical evolution 190–2
technical knowledge 184
technical objects 17, 128–9, 151, 164, 184, 189–200
technology 130
transduction 21–2
use 128–9
veritable relations 17, 19
slip resistance 123f–4f
Smith, Cyril Stanley 1, 2, 3, 142
Smith, Cyril Stanley and Hawthorne, John 72
Soane, John 40
　Tendring Hall 37
'Soft Office' (NOX) 87
sound insulation 118
Soundscape of Modernity, The (Thompson, Emily) 20
Specification (journal) 40–2f

INDEX

advertisements 60
proprietary specification 59, 60
timber naming 55–6
Yorke, F.R.S. 59, 60, 65
Specification for building a workshop (Mocatta, David) 107
'Specification for the Elfrida Rathbone School' (LCC). *See* Elfrida Rathbone School for the Educationally Subnormal
Specification for the Erection of a School house at Hammersmith (Fowler, Charles) 75
'Specification for the H. . .y Building' (Schuman Smith) 31, 49, 138–9
specification for the London Fever Hospital (Mocatta, David) 61, 75
Specification for Two houses on Marylebone Street (Mocatta, David) 54
specifications 6–10, 170–1
  architectural production 167–9
  building culture 8
  Burroughs, Brady 202
  changes over time 27, 29–32, 38
  claims by contractors 30
  clauses. *See* clauses
  collections of 40
  concrete 11
  descriptive 48, 49, 171
  Design and Build contract 30
  dimensions-led 31, 34, 35, 37–8, 51
  ends-oriented 31, 51
  examples 42
  Fore Street Bakehouse, dwelling and lofts 34–5
  form/matter schema. *See* form/matter schema
  forms of 187
  Foucault, Michel 32
  geography 7–8
  glass 137–8
  gross tendering 38–9
  H. . .y Building 31, 49, 138–9
  invention 201–2
  Knightly Brown, Andrew 28
  length of 30
  London Fever Hospital 61, 75

Marylebone Street houses 54
material properties 16
methods 47
mobilization 181, 182, 184–7
naming materials in 8–9
NBS. *See* NBS
Newgate Gaol and the Sessions House, Old Bailey. *See* Newgate Gaol, Old Bailey *and* Sessions House, Old Bailey
19th century expansion 40
open specifications 9, 31, 113–14, 171, 174
organization of 28–32
performance clauses. *See* performance clauses
performance criteria 28
performance. *See* performance specification
preliminary operations 13, 14
prescriptive 48, 49
process-based clauses *See* process-based clauses
product placement 202
progeny 43
proprietary 52–3, 59–65
redundant 177–8
RIBA 43–4
Schuman Smith 20, 28–9, 30, 48–50, 171
St James Square town house 31, 32–4
standard 31, 42–3
studies of 6–7
survival of 27
texture 186
trades based 28, 29–30, 31, 35–8, 40–1f, 51
transduction 23
transitional contracts 37
unconventional construction 175–9, 185
Specifications (Mocatta, David) 40, 42
Specifications for Practical Architecture (Bartholomew, Alfred) 40
Specifications in Detail (Macey, Frank) 42

Specifying Architecture: A Guide to Professional Practice (Gelder, John) 6, 47–8
speed bumps 143, 151
St James Burial Ground, Tottenham Court Road 71–2, 73
St James Square town house 31, 32–4
stand design 59–60
statements 29
static equilibrium 118
Stirling, James
   Leicester Engineering Building 168
Stock Orchard Street straw bale house (Sarah Wigglesworth Architects) 177–9, 180f, 181, 184–5
stone 120–2
   natural 49, 120–1, 172
straw bales 179, 180f
Sure Start, St Anne's Ward, Colchester (DSDHA) 56–7
Svenningsen Kajita, Heidi
   '(Im) Possible Instructions: Inscribing use-value in the architectural design process' 202
systems 153–6, 188
   constitutive seam 166–82
   informed material 156–60
   Simondon's 'complete system' 153–5, 159–66

Table 2/3 67–70, 71
technical activity 192–3
technical culture 192, 195, 196–7, 199
'Technical Culture and the Limits of Interaction: A Note on Simondon' (Toscano, Alberto) 196
technical debates 4
technical ensembles 192, 200
technical evolution 190–2, 199
technical knowledge 184
technical objects 16, 189–98 *see also* equipment
   Akrich, Madeleine 150
   evolution 190–2

Simondon, Gilbert 17, 128–9, 151, 164, 184, 189–200
technical operations 189, 190, 198–9 *see also* work
technics 12
technology 117, 130, 198 *see also* technical operations
Tecton 74
telephones 12, 190–1
Tendring Hall 37
texture 186
Theophilius 76
   Art of the Metalworker, The 85–6
   On Divers Arts 72, 85
Thompson, Emily
   Soundscape of Modernity, The 20
thought 22–3, 24
Thousand Plateaus, A (Deleuze, Gilles and Guattari, Félix) 90–1
3D printing 199
timber
   gopher wood 51
   implicit forms 92
   naming 51, 53–9
Timber Construction (publication) 57
tools 190, 195
Toscano, Alberto 161–2
   'Technical Culture and the Limits of Interaction: A Note on Simondon' 196
trade fairs 59–60
transduction 21–3, 24, 25
transductive thought 22–3, 24
transindividuation 189, 193
transitional contracts 37
translation concept 145, 151
truth to materials approach 167

unconventional construction 175–81, 184–5
UNICLASS (United Classification for the Construction Industry) 46
uniformity 105–6
use/usefulness 15, 16–17, 128–34, 138
   environmental performance 140–1
   Latour, Bruno 143–4
   performance-engineered materials 135, 142

User's Guide to Capitalism and Schizophrenia, A (Massumi, Brian) 110
USINA CTAH (Centro de Trabalhos para o Ambiente Habitado/ FACTORY – Work Centre for Living Space) 201

variety 66
veritable relations 17, 19
Vibrant Architecture (Armstrong, Rachel) 2
Vibrant Matter (Bennett, Jane) 2
visual intent 3, 173–4, 177
visual intent/performance 174
Vitruvius
 De Architectura 72
Voss, Daniela 191, 195–6, 197, 199

'Wall One' (Chandler, Alan and Pedreschi, Remo) 87–90, 198
Welsh Assembly Building (Richard Rogers and Partners) 149
Whitehead, Alfred North
 Process and Reality 157–8
wood. *See* timber
woodworking 110
work 193–7 *see also* labour
work-in-place approach 46
Worndl, Hans Peter
 Gucklhupf, Mondsee 57
Wrench, The (Levi, Primo) 85

Yaneva, Albena 149
Yorke, F.R.S. 59, 60, 65

www.ingramcontent.com/pod-product-compliance
Lightning Source LLC
Chambersburg PA
CBHW062128300426
44115CB00012BA/1847